Routledge Revivals

Chance and Error

Originally published in 1923 *Chance and Error* examines the vagaries of chance, and how this is the result of the interference of yes and no. The book basis its examination of chance on the idea of a two-sided coin. The book stipulates that contradictories are head and tail, or yes and no. When the coin is flipped in the air yes normally wins half of the trials, but this includes half of the half that normally go to no. Thus, normally in one quarter of the trials there is an interference of yes and no. From this the chance of any number of heads or tails can be easily calculated, and all results that are attained by more difficult mathematics are secured. The book uses this idea to examine interference of yes and no in everyday life and argues that this causes the variations in everything that goes on around us in nature and in our daily life. This book will be of interest to philosophers of logic, as well as mathematicians.

Chance and Error

The Theory of Evolution

By Marsh Hopkins

Routledge
Taylor & Francis Group

First published in 1923
by Kegan Paul, Trench, Trubner & Co., Ltd.

This edition first published in 2019 by Routledge
2 Park Square, Milton Park, Abingdon, Oxon, OX14 4RN
and by Routledge
52 Vanderbilt Avenue, New York, NY 10017

Routledge is an imprint of the Taylor & Francis Group, an informa business

Publisher's Note

The publisher has gone to great lengths to ensure the quality of this reprint but points out that some imperfections in the original copies may be apparent.

Disclaimer

The publisher has made every effort to trace copyright holders and welcomes correspondence from those they have been unable to contact.

A Library of Congress record exists under LCCN: 24004115

ISBN 13: 978-0-367-26616-5 (hbk)
ISBN 13: 978-0-429-29420-4 (ebk)

CHANCE AND ERROR

CHANCE AND ERROR

The Theory of Evolution

BY

MARSH HOPKINS

B.A.Sc., M.E.I.C., D.L.S.

"So careful of the type she seems
So careless of the single life."

LONDON

KEGAN PAUL, TRENCH, TRUBNER & Co., LTD.

NEW YORK: E. P. DUTTON & Co.

1923

PREFACE

THIS little book shows that the vagaries of chance are the result of the interference of yes and no. A coin has two similar sides of which one is head and the other tail. The contradictories are head and tail, or yes and no. When the coin is flipped in the air yes normally wins half of the trials, but this includes half of the half that normally go to no. Thus normally in one quarter of the trials there is an interference of yes and no. From this the chance of any number of heads or tails can be easily calculated, and all the results that are attained by more difficult mathematics are secured.

Several formulæ are worked out that are remarkably simple and precise and can be used mentally and without the use of any Tables.

It was written with the object of extending the usefulness of this very important subject to those whose knowledge of mathematics is limited.

The subject took such a hold of me that for a number of years I constantly worked at it in my sleep.

The interference of yes and no causes the variations in everything that goes on around us in nature and in our daily life, and comes home to the bosoms of men more closely than any other subject.

St. Paul, Alberta, Jan. 3, 1923. THE AUTHOR.

CONTENTS

Chapter I

PROPOSITIONS

Definition 1

Chance is unknown cause.

Definition 2

A chance of anything is one of the ways in which it can happen or fail when all the ways seem equally likely.

Definition 3

The chance of anything is the fraction of all its chances that is favourable to its happening.

It is also the number of ways in which it can happen when the sum of its chances is equal to 1.

Definition 4

A combination is any collection of things, and has no regard to how the things are arranged.

It is also called a selection or parcel.

A blank combination contains none of the things.

Definition 5

An arrangement is any one of the orders in which the things in a combination can be arranged.

It is also called a permutation or group.

Definition 6

When the possible arrangements of the things in a combination are not to be considered the things are regarded as all alike, and are said to be of the same sort. But if we think of them as being arranged they are said to be different.

Any number of things that are regarded as all alike can be arranged in but one order.

Definition 7

An element of anything is any piece of it that may be taken at any one time.

Definition 8

Yes and no at the same time are contradictories.

Definition 9

Lines that tend to contradictories are called contradictory lines.

Definition 10

The component in any direction of any line is the projection of the line on that direction.

Definition 11

An error is the result of chance less the expectation or true value.

Definition 12

The normal error in any set of conditions is such that it may be substituted for each error made under that set of conditions without changing the chance that all the errors will occur.

Definition 13

The weight of any set of conditions is equal to the number of times that an observation made under that set of conditions must be taken to produce the same number of chances of a pair of contradictories as a single observation made under a set of conditions whose weight is 1.

And the weight of any observation is equal to the weight of the set of conditions under which it is made.

Definition 14

An element of anything is very small when its magnitude may be disregarded.

Definition 15

Either a win or a failure is called an event.

Definition 16

Anything that tends to occur but cannot occur is imaginary.

Definition 17

Any line tending to an imaginary thing is called an imaginary line.

Definition 18

Any angle is naturally measured by its circular arc over the radius.

Definition 19

The Natural chance of anything is the true chance when a unit is a very small element of the thing.

Axiom 1

Contradictories cannot have a common element.

Axiom 2

The real and the imaginary cannot have a common element.

Axiom 3

An error is the result of yes and no tending to occur at the same time.

Axiom 4

Any succession of a contradictory can be represented in magnitude and direction by a straight line.

Axiom 5

If two lines are at right angles to each other neither has a component in the direction of the other.

Axiom 6

If two lines are not at right angles to each other either line has a component in the direction of the other.

Axiom 7

The sum of the squares of the two sides of a right angled triangle is equal to the square of the hypotenuse.

Axiom 8

If one thing can be done in r ways, and, after it is done, another thing can be done in s ways, the second thing can be done in s ways with each of the r ways in which the first thing can be done, and the two things can be done in this order in rs ways. And, in general, if any number of things can be done in succession in r, s, t, and so on, ways, respectively, all the things can be done, in this order, in rst and so on ways.

Axiom 9

If there is a combination containing r things any of them may be placed first, and, after one is placed, any of the remaining $r-1$ things may be placed second, and, after 2 are placed, any of the remaining $r-2$ things may be placed third, and so on. And they can all be arranged in

$$r(r-1)(r-2)(r-3)\ldots 1 = 1.2.3.\ldots.r = \underline{|r} \text{ ways}$$

Axiom 10

All combinations that can be made in the same number of equally likely ways are equally likely.

Examples

Example 1. A and B met by chance means that they met by unknown cause. There was cause for their both being there, but it was unknown to them. If it had been known to them they could not say that they met by chance, by Definition 1.

If A had 2 chances of being there and 998 chances of not being there, and B had 6 chances of being there and 91 chances of not being there, what is the chance that they both would be there? What is the chance that A alone would be there?

A could be there in 2 ways and not there in 998 ways, and B could be there in 6 ways and not there in 91 ways, when all these ways seem equally likely, by Definition 2.

Hence the chance that A would be there is $p_1 = \frac{2}{1000}$ by Definition 3, and the chance that B would be there is $p_2 = \frac{6}{97}$, by Definition 3.

A and B both could be there in $2 \times 6 = 12$ ways, by Axiom 8.

A and B both could be there and not there in $(2 + 998)(6 + 91) = 1000 \times 97$ ways, by Axiom 8.

Hence the chance that A and B both would be there is $\frac{2 \times 6}{(2 + 998)(6 + 91)} = \frac{2}{1000} \times \frac{6}{97} = p_1 p_2 = \cdot 00012$, by Definition 3.

Hence the chance that A alone would be there is $p_1 - p_1 p_2 = \frac{2}{1000} - \frac{12}{97000} = \frac{182}{97000} = \cdot 00188$, by Definition 3.

The ways may seem equally likely to one person and not equally likely to another. Since chance is unknown cause if one person knows more about anything than another person, the chance of its happening is not the same to the two persons. And if one knows all about anything there is no chance at all to him with regard to it, by Definition 1.

Example 2. A bag of flour is composed of very small elements, each speck of flour being an element of the bag, by Definition 7.

Example 3. A combination of the letters *a, b* and *c* can be arranged in the different groups, *abc, bca, cab, cba, acb* and *bac*, each of which is a different group, permutation or arrangement, but they are all the same combination, by Definitions 4 and 5.

Example 4. Are to win in one trial and to fail in another trial contradictories ?

Contradictories are only in the same trial or the same game, that is, at the same time, by Definition 8.

Example 5. What is the value of the component of a line in the direction of the line ?

It is equal to the line, by Definition 10.

Example 6. What is the value of a component in a direction at right angles to the direction of the line ?

It is 0, by Definition 10.

Example 7. If in a raffle for $10 there are 10 tickets sold for $1 each and a man buys 2 tickets, what is his expectation ?

The chance that he will win $8 is $\frac{1}{5}$, and the chance that he will lose $2 is $\frac{4}{5}$, by Def. 3.

Hence his expectation is

$\frac{1}{5}$ of \$8 — $\frac{4}{5}$ of \$2 = 0 by Def. 11 and Axiom 3.

Example 8. If he has a chance of $\frac{1}{5}$ to win $5 and a chance of $\frac{4}{5}$ to lose $1, what is his expectation ?

It is $\frac{1}{5}$ of \$5 — $\frac{4}{5}$ of \$1 = 20 cents.

Example 9. Can to win and to win and fail in the same trial have a common element ?

They cannot, because to win is real and to win and fail in the same trial is imaginary, by Definition 16 and Axiom 2.

A coin flipped in the air tends to be head one half the times and tail one half the times, and both head and tail $\frac{1}{2}$ of $\frac{1}{2} = \frac{1}{4}$ of the times, by Definition 3. But it cannot be both head and tail in the same trial. Hence both head and tail in the same trial is imaginary, by Definition 16.

If there is no error it will be head the same number of times that it is tail, by Definition 3 and 11.

Example 10. If two contradictory successions each equal to 4 both tend to occur what is the error ?

The error is $\sqrt{4^2 + 4^2} = \sqrt{32}$. by Axioms 3, 5, and 7.

Example 11. If there is a succession of 6 things what is the magnitude of the succession ?

It is 6, by Axiom 4.

Example 12. If two horses are pulling on a load in directions at right angles to each other, what is the effect of one on the other ?

It is 0, by Definition 10, and Axiom 5.

Example 13. If they are pulling in directions that are not at right angles to each other, what is the effect of one on the other ?

It is the component of its force in the direction in which the other horse is pulling, by Definition 10.

Example 14 If the sides of a right angled triangle are 3 and 4, what is the hypotenuse ?

It is $\sqrt{3^2 + 4^2} = 5$, by Axiom 7.

Example 15. If a man has 4 knives and 5 forks to choose from, in how many ways can he take a different combination of a knife and a fork.

In $4 \times 5 = 20$ ways, by Axiom 8 and Definitions 4 and 6.

Any fork can be taken with each of the 4 knives.

Hence a different combination of a knife and fork can be taken in $4 \times 5 = 20$ ways, by Definition 4.

A different group of a knife and fork can be taken in $20 \times \underline{|2} = 40$ ways, by Axiom 9 and Definition 5.

Example 16. If he has 4 knives and 5 forks and 6 spoons to choose from, in how many ways can he take a different combination of a knife, fork and spoon?

In $4 \times 5 \times 6 = 120$ ways, by Axiom 8.

Any spoon can be taken with each of the 20 different combinations of a knife and fork.

A different group of them can be taken in

$$120 \times \underline{|3} = 720 \text{ ways.}$$

Example 17. In how many ways can 5 boys be arranged in a class?

In $\underline{|5} = 5.4.3.2.1. = 120$ ways, by Axiom 9.

Example 18. In how many ways can a knife, fork and spoon be arranged?

In $\underline{|3} = 6$ ways.

Example 19. In how many different orders can a knife, fork, spoon and cup be taken?

In $\underline{|4} = 24$ orders, by Axiom 9.

Example 20. Show that if there are n things, of which r are alike of one sort, and s alike of another sort, and t alike of another sort, etc., the number of ways in which the n things can be arranged is $\dfrac{\underline{|n}}{\underline{|r}\ \underline{|s}\ \underline{|t}\text{ etc.}}$.

If they are all different they can be arranged in $\underline{|n}$ ways, by Axiom 9.

And r different things can be arranged in $\underline{|r}$ different orders, and s different things in $\underline{|s}$ different orders, and t different things in $\underline{|t}$ different orders, and so on, by Axiom 9.

But if the r things are of the same sort they can be arranged in only one way, by Definition 6, and the s things of one sort can be arranged in only one way, and the t things of one sort can be arranged in only one way.

Hence when r things are of one sort and s things of another sort and t things of another sort, and so on, they can be arranged in $\dfrac{1}{\underline{|r}\ \underline{|s}\ \underline{|t}\text{, etc.}}$ as many ways as when they are all different, by Axiom 8.

But when they are all different the n things can be arranged in $\underline{|n}$ ways.

Hence when r things are of one sort and s things of another sort, and t things of another sort, and so on, the n things can be arranged in $\dfrac{\underline{|n}}{\underline{|r}\ \underline{|s}\ \underline{|t}\text{ etc.}}$ ways.

Example 21. In how many ways can the letters *aabc* be arranged? And in how many ways can the letters of the words Chance and Error be arranged?

The first in $\dfrac{\underline{|n}}{\underline{|r}\ \underline{|s}\ \underline{|t}} = \dfrac{\underline{|4}}{\underline{|2}\ \underline{|1}\ \underline{|1}} = \dfrac{\underline{|4}}{\underline{|2}} = 12$ ways.

And in the words Chance and Error c occurs twice, h once, a twice, e twice, n twice, d once, r three times and o once.

Hence the number of ways in which the letters can be arranged is $\dfrac{\underline{|n}}{\underline{|r}\ \underline{|s}\ \underline{|t}\text{ etc.}} = \dfrac{\underline{|14}}{(\underline{|2})^4\ \underline{|3}} = 908107200.$

Example 22. If anything is composed of a single element, what is the difference between the Natural chance and the Common or true chance ?

The Natural chance assumes the series to be so large that a unit is a very small element of it.

Hence if the error is 3 feet a foot is not likely a very small element of it, but 3 feet is the same as 36 inches or 3600 hundredths of an inch.

Hence a unit can always be made a very small element when the error is expressed in length, very small depending on the precision required, by Definition 14. We can take a unit as small as we like.

PROPOSITION 1

The number of ways in which a different combination containing r things can be made from n things each composed of m equal elements is

$$n\left(n-\frac{1}{m}\right)\left(n-\frac{2}{m}\right)\left(n-\frac{3}{m}\right)\cdots\cdots\left(n-\frac{r-1}{m}\right)\frac{1}{r!}=$$

$$\frac{(mn)!}{r!\,(mn-r)!}\left(\frac{1}{m}\right)^r.$$

And the elements need not be equal so long as they are all very small.

Since there are n things each composed of m equal elements there are mn equal elements altogether.

And an element can be taken at any one time, by Definition 7.

Hence the first element can be taken in mn ways, since there are mn elements to choose from.

After this is done there are $mn-1$ elements left.

Hence the second element can be taken in $mn-1$ ways.

Hence the first two elements can be taken in $mn\,(mn-1)$ ways, by Axiom 8.

And the two elements can be arranged as first and second in $\underline{|2}$ ways, by Axiom 9.

Hence a different combination of 2 elements can be made in $\frac{mn\,(mn-1)}{\underline{|2}}$ ways, by Definition 4.

After 2 elements are taken there are $mn - 2$ left.

Hence the third element can be taken in $mn - 2$ ways.

Hence the first, second and third elements can be taken in $mn\,(mn-1)\,(mn-2)$ ways, by Axiom 8.

And these three elements can be arranged as first, second and third in $\underline{|3}$ ways, by Axiom 9.

Hence a different combination of 3 elements can be made in $\frac{mn\,(mn-1)\,(mn-2)}{\underline{|3}}$ ways, by Definition 4.

Similarly a different combination of r elements can be made in

$$\frac{mn\,(mn-1)\,(mn-2)\,(mn-3)\,..\,(mn-\overline{r-1})}{\underline{|r}}$$

$$= \frac{\underline{|mn}}{\underline{|r}\ \underline{|mn-r}} \text{ ways.}$$

And since there are m elements in a thing an element can be taken in m times as many ways as a thing.

Hence r elements can be taken in m^r times as many ways as r things, by Axiom 8.

Hence r things can be taken in $\frac{\underline{|mn}}{\underline{|mn-r}}\left(\frac{1}{m}\right)^r$ ways.

Hence a different combination of r things can be made

$$\frac{\underline{|mn}}{\underline{|r}\ \underline{|mn-r}}\left(\frac{1}{m}\right)^r =$$

$$\frac{n\,(n-\frac{1}{m})\,(n-\frac{2}{m})\,\ldots\,(n-\frac{r-1}{m})}{\underline{|r}} \text{ ways.}$$

And the elements need not be equal so long as they are all very small, by Definition 14.

Hence the Proposition is true.

Example 23. If there are 4 bags of marbles, in how many ways can a different combination, each containing 2 bags, be made from the 4 bags ?

If fractions are not allowed the first bag can be taken in 4 ways, and the second in 3 ways.

And these 2 bags can be arranged as first and second in $|\underline{2}$ ways, by Axiom 9.

Hence a different combination of 2 bags can be made from the 4 bags in $\frac{4 \cdot 3}{|\underline{2}} = 6$ ways, by Definition 4.

But if fractions are allowed, and an element is $\frac{1}{5}$ of a bag, we may take one element at any one time instead of a whole bag, by Definition 7.

Hence a different combination containing 2 bags can be made in $\frac{4\,(4 - \frac{1}{5})}{|\underline{2}} = 7 \cdot 6$ ways.

If the elements are all very small the number of ways is $\frac{4 \times 4}{|\underline{2}} = 8$, by Definition 14.

Example 24. If each bag is composed of a single element, in how many ways can a different combination containing 2 bags be made from the 4 bags ? That is, what is the number of combinations of 4 things 2 at a time ?

It is $\frac{4 \cdot 3}{|\underline{2}} = 6$.

Proposition 2

The number of ways in which a different combination of things can be made from n things, each composed of very small elements is

$$e^n = 1 + n + \frac{n^2}{\underline{|2}} + \frac{n^3}{\underline{|3}} + \frac{n^4}{\underline{|4}} + \text{etc.}$$

where $e = 2{\cdot}718281828 +$.

And if each thing is composed of a single element the number of different combinations is

$$2^n = 1 + n + \frac{n\,(n-1)}{\underline{|2}} + \frac{n(n-1)\;(n-2)}{\underline{|3}} + \text{etc.} =$$

$$\Sigma \frac{\underline{|n}}{\underline{|r}\;\underline{|n-r}}.$$

By Proposition 1 the number of ways in which a different combination containing r things can be made is

$$\frac{n\left(n-\frac{1}{m}\right)\left(n-\frac{2}{m}\right)\cdots\left(n-\frac{r-1}{m}\right)}{\underline{|r}}$$

$$= \frac{\underline{|mn}}{\underline{|r}\;\underline{|mn-r}}\left(\frac{1}{m}\right)^r$$

in which r may successively take the values 0, 1, 2, 3, and so on, by Definition 4.

If there is only one element $n = \frac{1}{m}$.

When $r = 0$, $\dfrac{\underline{|mn}}{\underline{|r}\;\underline{|mn-r}}\left(\dfrac{1}{m}\right)^r = 1.$

when $r = 1$ it is equal to $\frac{1}{m}$.

Hence if there is only one element the number of ways in which a different combination can be made is $1 + \frac{1}{m}$.

And one of these different combinations is blank.

And after these $1 + \frac{1}{m}$ different combinations have been made $1 + \frac{1}{m}$ different combinations can be made from any other element.

Hence, since there are mn elements, the total number of ways in which a different combination can be made is $(1 + \frac{1}{m})^{mn}$, by Axiom 8.

Hence $(1 + \frac{1}{m})^{mn} = \Sigma \frac{\underline{|mn}}{\underline{|r}\ \underline{|mn - r}} \left(\frac{1}{m}\right)^r$,

where r successively takes the values 0, 1, 2, 3 and so on, and the results all added.

Hence the total number of different combinations of things is

$$(1 + \tfrac{1}{m})^{mn} = 1 + n + \frac{n\,(n - \frac{1}{m})}{\underline{|2}} + \frac{n\,(n - \frac{1}{m})\,(n - \frac{2}{m})}{\underline{|3}} + \text{etc.}$$

Hence if each thing is composed of a single element the total number of combinations is

$$(1 + 1)^n = 2^n = 1 + n + \frac{n\,(n - 1)}{\underline{|2}} + \frac{n\,(n-1)\,(n-2)}{\underline{|3}} + \text{etc.}$$

$$= \Sigma \frac{\underline{|n}}{\underline{|r}\ \underline{|n - r}}.$$

As $\frac{1}{m}$ becomes smaller the numerator becomes larger. And this change is continuous right down to $\frac{1}{m} = 0$, which gives the maximum value of the number of combinations.

If each thing is composed of an infinite number of elements $\frac{1}{m} = 0$, and when $n = 1$ the expression becomes

$$(1 + 0)^{\frac{1}{0}} = 1 + 1 + \frac{1}{\underline{|2}} + \frac{1}{\underline{|3}} + \frac{1}{\underline{|4}} + \text{etc.} = 2{\cdot}718281828 + .$$

This is usually denoted by e.

And the number of ways in which a different combination of things can be made from n things is

$$\left\{(1+0)^{\frac{1}{0}}\right\}^n = e^n = 1 + n + \frac{n^2}{\lfloor 2} + \frac{n^3}{\lfloor 3} + \text{etc.}$$

Hence the Proposition is true.

Example 25. What is the value of e^{-x} when x is very small?

$$e^{-x} = \left(e^{-\frac{x}{2}}\right)^2 = \frac{e^{-\frac{x}{2}}}{e^{+\frac{x}{2}}} = \frac{1 - \frac{x}{2} + \frac{x^2}{8} - \text{etc.}}{1 + \frac{x}{2} + \frac{x^2}{8} + \text{etc.}}$$

$$= \frac{1 - \frac{x}{2}}{1 + \frac{x}{2}} \text{ if } x \text{ is very small, by Definition 14,}$$

$$= \frac{2 - x}{2 + x}.$$

Example 26. What is the value of $e^{-\cdot 137}$?

$$e^{-\cdot 137} = \frac{2 - \cdot 137}{2 + \cdot 137} = \frac{1.863}{2{\cdot}137}, \text{ very nearly.}$$

Example 27. What is the value of $\left(\frac{99}{100}\right)^{10}$?

$$\left(\frac{99}{100}\right)^{10} = \left(\frac{2 - \frac{1}{99.5}}{2 + \frac{1}{99{\cdot}5}}\right)^{10} = \left(e^{-\frac{1}{99\cdot 5}}\right)^{10} = e^{-\frac{10}{99\cdot 5}}$$

$$= \frac{2 - \frac{10}{99\cdot 5}}{2 + \frac{10}{99\cdot 5}} = \frac{199 - 10}{199 + 10} = \frac{189}{209}, \text{ very nearly.}$$

Example 28. What is the value of $\left(\frac{a}{b}\right)^n$ when $n\,(a - b)$ is very small?

It is $\dfrac{a + b + n\,(a - b)}{a + b - n\,(a - b)} = e^{2n\,\frac{a-b}{a+b}}$, by Definition 14.

Example 29. If the expectation is that an earthquake will occur once a year, and it is equally likely to occur in any instant, what is the chance that it will not occur at all in four years? What is the chance that it will occur 12 times in any four years.

If the elements are all very small a different combination containing r years can be made from 4 years in

$\frac{4^r}{\underline{|r}}$ ways, by Proposition 1.

And the total number of combinations of years is e^4, by Proposition 2.

And all these combinations are equally likely, by Axiom 10.

Hence the chance that an earthquake will occur r times in 4 years is $\frac{\frac{4^r}{\underline{|r}}}{e^4} = \frac{4^r}{\underline{|r}} e^{-4}$, by Def 3 & Prop. 2.

Hence the chance that there will be no earthquakes at all in the 4 years is $e^{-4} = \cdot 0183$, by Proposition 2.

And the chance of 4 earthquakes is $\frac{4^4}{\underline{|4}} e^{-4} = \cdot 0183 \times \frac{32}{3} = \cdot 1952$, by Proposition 2.

And the chance of 12 earthquakes is $\frac{4^{12}}{\underline{|12}} e^{-4} = \cdot 00064$ by Proposition 2.

Example 30. In how many ways can a different combination of bags be made from the elements of 3 bags of flour?

The elements of a bag of flour are all very small pieces of the bag, by Definition 7.

Hence a different combination of bags can be made from the elements of 3 bags of flour in $e^3 = 20 \cdot 0855$ ways, by Proposition 2.

And one of these different combinations is blank.

Example 31. In how many ways can a different combination be made from 3 things each composed of a single element ?

In $2^3 = 8$ ways, by Proposition 2.

And one of these is blank.

Example 32. In how many ways can a different combination, each containing 2 things, be made from 3 things not subdivided ?

In $\frac{\underline{|n}}{\underline{|r}\,\underline{|n-r}} = \frac{\underline{|3}}{\underline{|2}\,\underline{|1}} = 3$ ways, by Proposition 2.

Example 33. In how many ways can a different combination containing 4 things be made from 6 elemental things ? Or what is the number of combinations of 6 things 4 at a time ?

It is $\frac{\underline{|6}}{\underline{|4}\,\underline{|2}} = 15.$

Example 34. In how many ways can 6 boys be divided up into 2 combinations of 3 boys each ?

The first 3 boys can be selected in $\frac{\underline{|6}}{\underline{|3}\,\underline{|3}} = 20$ ways.

And after the first 3 are selected there are 3 boys left from which to select the second 3 boys.

Hence the second 3 boys can be selected in $\frac{\underline{|3}}{\underline{|3}\,\underline{|0}} =$ 1 way,

Hence the first and second 3 can be selected in $\frac{\underline{|6}}{\underline{|3}\,\underline{|3}} \times \frac{\underline{|3}}{\underline{|3}\,\underline{|0}} = \frac{\underline{|6}}{\underline{|3}\,\underline{|3}} = 20$ ways, by Axiom 8.

$\underline{|0} = 1.$

But since each of these two combinations may successively contain the same 3 boys, first as the first 3 and then as the second 3, the number of different combinations is only $\frac{1}{\underline{|2}}$ of this, by Definition 4 and Axiom 9.

Hence the number of ways in which 6 boys can be divided into 2 combinations each containing 3 boys is

$$\frac{\underline{|6}}{\underline{|3}\ \underline{|3}\ \underline{|2}} = 10.$$

Example 35. In how many ways can 8 boys be divided up into 4 combinations each containing 2 boys ?

The first 2 boys can be selected in $\frac{\underline{|8}}{\underline{|2}\ \underline{|6}} = 28$ ways.

There are 6 boys left from which to select the second 2.

Hence this can be done in $\frac{\underline{|6}}{\underline{|2}\ \underline{|4}} = 15$ ways.

There are 4 boys left from which to select the third 2 boys.

Hence this can be done in $\frac{\underline{|4}}{\underline{|2}\ \underline{|2}} = 6$ ways.

And the fourth 2 boys can be selected in $\frac{\underline{|2}}{\underline{|2}\ \underline{|0}}$ $= 1$ way.

Hence the 4 combinations each containing 2 boys can be made and arranged as the first, second, third and fourth 2 in

$$\frac{\underline{|8}}{\underline{|2}\ \underline{|6}} \times \frac{\underline{|6}}{\underline{|2}\ \underline{|4}} \times \frac{\underline{|4}}{\underline{|2}\ \underline{|2}} \times \frac{\underline{|2}}{\underline{|2}\ \underline{|0}} = \frac{\underline{|8}}{\underline{|2}\ \underline{|2}\ \underline{|2}\ \underline{|2}}$$

$$= \frac{\underline{|8}}{(\underline{|2})^4} \text{ ways.}$$

And the 4 combinations can be arranged in $\underline{|4}$ ways, by Axiom 9.

Hence they can be made and not arranged in $\frac{\underline{|8}}{(\underline{|2})^4 \underline{|4}}$ ways, which is the number of combinations, by Definition 4.

And in general n things can be divided up into different combinations containing r, s, t, and so on, things, respectively, in $\frac{\underline{|n}}{\underline{|r}\ \underline{|s}\ \underline{|t}\ \text{etc.}}$ ways, if r, s, t, and so on, are all different.

And the n things can be given r, s, t, and so on, respectively, to different persons in $\frac{\underline{|n}}{\underline{|r}\ \underline{|s}\ \underline{|t}\ \text{etc.}}$ ways, whether r, s, t, and so on, are all different or not, because giving them to different persons makes them different, by Definition 6.

Example 36. In how many ways can 9 boys be divided up into four different combinations of 1, 1, 3 and 4 boys, respectively?

In $\frac{\underline{|9}}{(\underline{|1})^2\ \underline{|2}\ \underline{|3}\ \underline{|4}} = 1260$ ways.

Example 37. In how many ways can 11 boys be divided up into 5 different combinations containing, 1, 1, 3, 3, 3, respectively?

In $\frac{\underline{|11}}{(\underline{|1})^2\ \underline{|2}\ (\underline{|3})^3\ \underline{|3}} = 15{,}400$ ways.

Example 38. In how many ways can a selection of 5 cards be made from a pack of 52 cards?

In $\frac{\underline{|52}}{\underline{|5}\ \underline{|47}}$ ways.

Example 39. In how many ways can 3 of a kind and 2 other cards be selected from a pack ?

Since there are 13 kinds, one kind can be selected in 13 ways.

And when this is done 3 cards can be selected from this kind containing 4 cards in $\frac{4!}{3!\,1!} = 4$ ways.

And when this is done 2 cards can be selected from the 48 cards of the remaining 12 kinds in $\frac{48!}{2!\,46!}$ ways.

Hence the hand of 5 cards can be selected in $13 \times 4 \times \frac{48!}{2!\,46!}$ ways, by Axiom 8.

Example 40. If the 2 cards are to be of 2 different kinds we can select the 2 kinds from the remaining 12 kinds in $\frac{12!}{2!\,10!}$ ways.

And when this is done we can select 1 card from each of the 2 kinds in 4^2 ways, by Axiom 8.

Hence the hand of 5 cards can be selected in $13 \times 4 \times \frac{12!}{2!\,10!} \times 4^2$ ways.

Example 41. If the 2 cards are to be of the same kind this kind can be selected from the remaining 12 kinds in 12 ways.

And when this is done 2 cards can be selected from this kind in $\frac{4!}{2!\,2!} = 6$ ways.

Hence the hand of 3 of a kind and a pair can be selected in $13 \times 4 \times 12 \times 6$ ways.

And a hand of 5 cards can be selected in $\frac{\lfloor 52}{\lfloor 5 \lfloor 47}$ ways.

Example 42. What is the chance of 3 of a kind and a pair in a hand of 5 cards in the first draw?

It is $\dfrac{13 \times 4 \times 12 \times 6}{\dfrac{\lfloor 52}{\lfloor 5 \lfloor 47}} = \dfrac{6}{4165} = \cdot 00144$, by Def. 3.

Example 43. What is the chance of 3 of a kind and 2 odd cards?

$$\text{It is } \dfrac{13 \times 4 \times \dfrac{\lfloor 12}{\lfloor 2 \lfloor 10} \times 4^2}{\dfrac{\lfloor 52}{\lfloor 5 \lfloor 47}} = \cdot 02113.$$

Example 44. What is the chance of 2 pairs and an odd card?

The 5 cards must be taken from 3 different kinds.

Three kinds can be selected from 13 kinds in $\frac{\lfloor 13}{\lfloor 3 \lfloor 10}$ ways.

And when this is done 2 kinds can be selected from these 3 kinds in $\frac{\lfloor 3}{\lfloor 2 \lfloor 1}$ ways.

And a pair can be selected from each of these 2 kinds in $\left(\frac{\lfloor 4}{\lfloor 2 \lfloor 2}\right)^2$ ways, by Axiom 8.

And the single card can be selected from the other kind in 4 ways.

Hence the hand of 5 cards can be selected in $\frac{\lfloor 13}{\lfloor 3 \lfloor 10} \times 3 \times 6^2 \times 4$ ways, by Axiom 8.

Hence the chance of 2 pairs and an odd card in a hand of 5 cards is

$$\frac{\dfrac{\lfloor 13}{\lfloor 3\ \lfloor 10} \times 3 \times 6^2 \times 4}{\dfrac{\lfloor 52}{\lfloor 5\ \lfloor 47}} = \frac{198}{4165} = \cdot 04754.$$

Example 45. What is the chance of a single pair and 3 odd cards ?

The hand of 5 cards must be taken from 4 different kinds.

Four kinds can be selected from 13 kinds in $\dfrac{\lfloor 13}{\lfloor 4\ \lfloor 9}$ ways.

And when this is done one kind can be selected from these 4 kinds in 4 ways.

And a pair can be selected from this kind in $\dfrac{\lfloor 4}{\lfloor 2\ \lfloor 2}$ ways.

And a card can be selected from each of the remaining 3 kinds in 4^3 ways, by Axiom 8.

Hence the hand can be selected in

$$\frac{\lfloor 13}{\lfloor 4\ \lfloor 9} \times 4 \times \frac{\lfloor 4}{\lfloor 2\ \lfloor 2} \times 4^3 \text{ ways.}$$

Hence the chance of a pair and 3 odd cards is

$$\frac{\dfrac{\lfloor 13}{\lfloor 4\ \lfloor 9} \times 4^4 \times \dfrac{\lfloor 4}{\lfloor 2\ \lfloor 2}}{\dfrac{\lfloor 52}{\lfloor 5\ \lfloor 47}} = \frac{1056}{2499} = \cdot 42257.$$

Example 46. What is the chance that the 5 cards will all be of the same suit ?

One suit can be selected from 4 suits in 4 ways.

And when this is done 5 cards can be selected from this suit of 13 cards in $\frac{\underline{|13}}{\underline{|5}\;\underline{|8}}$ ways.

Hence the chance of a flush is

$$\frac{4 \times \dfrac{\underline{|13}}{\underline{|5}\;\underline{|8}}}{\dfrac{\underline{|52}}{\underline{|5}\;\underline{|47}}} = \frac{33}{16{,}660} = \cdot 00110.$$

Example 47. In how many ways can a first and second baseball 9 be chosen from 18 players ?

In $\frac{\underline{|18}}{\underline{|9}\;\underline{|9}}$ ways.

Example 48. In how many ways can 2 baseball nines be chosen from 18 players ?

In $\frac{\underline{|18}}{\underline{|9}\;\underline{|9}\;\underline{|2}}$ ways.

Example 49. In how many ways can 3 different nines be chosen from 27 players ?

In $\frac{\underline{|27}}{\underline{|9}\;\underline{|9}\;\underline{|9}\;\underline{|3}}$ ways.

Example 50. In how many ways can a different combination be made from 18 men ?

In $2^{18} = 262141$ ways, by Proposition 2.

One of these is blank.

Hence if there is to be no blank the number of ways is $2^{18} - 1 = 262140$.

Example 51. If there are 3 news boys in how many ways can A buy 2 papers from them ?

He can buy one paper from any of the boys. And when this is done he can buy the other paper from any of the boys.

Hence he can buy the 2 papers from the 3 boys in $3 \times 3 = 3^2$ ways, by Axiom 8.

Example 52. If he can buy only 1 paper from the same boy, in how many ways can he buy 2 papers from 3 boys ?

He can buy the first from any of the 3 boys. And when this is done he can buy the second from either of the remaining boys.

Hence he can buy the 2 papers from 3 boys in $3 \times 2 = 6$ ways, by Axiom 8.

Example 53. If he has to buy both papers from the same boy, in how many ways can he buy 2 papers from 3 boys ?

He can buy the 2 papers from any of the boys.

Hence he can buy the 2 papers in 3 ways.

Example 54. What is the chance of drawing an ace and a king from a pack of 52 cards in the first draw ?

The number of ways in which a parcel of 2 cards can be made out of 52 cards is $\dfrac{|\underline{52}}{|\underline{2}\;|\underline{50}}$.

There are 4 aces and 4 kings in a pack of 52 cards.

And each ace can form a parcel of an ace and king with each king, making $4 \times 4 = 16$ different parcels.

Hence the chance of a parcel containing an ace and king is $\dfrac{16}{26 \times 51}$.

Example 55. What is the chance of drawing a king at the first draw ?

There are 4 kings, and 52 cards altogether.

Hence the chance of a king at the first draw is $\frac{4}{52}$, by Definition 3.

Example 56. What is the chance of drawing a king at the first and queen at the second draw ?

There are 52 cards to draw a king from, giving a chance of $\frac{4}{52}$. And since one card is drawn there are 51 left from which to draw a queen.

If the first was a king the chance of drawing a queen from the remaining 51 cards is $\frac{4}{51}$.

The chance that the first was a king is $\frac{4}{52}$.

Hence the chance that both the first was a king and the second a queen is $\frac{4}{52} \times \frac{4}{51} = \frac{16}{52 \times 51}$, by Definition 3.

The chance of drawing a queen first and a king second is the same as that of drawing a king first and a queen second.

Hence the chance of drawing either a king first and a queen second or a queen first and king second, that is, the chance of drawing a king and queen in the first 2 draws is $\frac{16}{52 \times 51} + \frac{16}{52 \times 51} = \frac{16}{26 \times 51}$, by Definition 3.

Example 57. What is the chance of 4 aces in a hand of five cards ?

There are 4 aces in a pack of cards.

Hence 4 aces can be selected in only 1 way. There are 48 cards that are not ace.

Hence the odd card can be selected in 48 ways.

Hence the hand can be selected in $1 \times 48 = 48$ ways.

And a hand of 5 cards can be selected from the 52 cards of the pack in $\dfrac{52!}{5!\,47!}$ ways.

Hence the chance of 4 aces in a hand of 5 cards is

$$\frac{48}{\dfrac{52!}{5!\,47!}} = \frac{5!\,48!}{52!}.$$

Example 58. What is the chance of 4 of a kind ?

It is $\dfrac{5!\,48!}{52!} \times 13$, since there are 13 kinds in a pack of cards.

Example 59. How many dominoes are there in a set made from the 7 numbers from 0 to 6 if doubles are not allowed ?

The number of combinations of 7 things, 2 at a time, is $\dfrac{n\,(n-1)}{2!} = \dfrac{7 \cdot 6}{2} = 21$, which is the answer, by Proposition 2.

Example 60. In how many ways can a different combination be made from 3 kinds of flowers if repetitions of any kind are not allowed ?

Including a blank combination it is

$$2^n = 1 + n + \frac{n\,(n-1)}{2!} + \frac{n\,(n-1)\,(n-2)}{3!} + \text{etc.}$$

$$= 2^3 = 1 + 3 + \frac{3 \cdot 2}{2} + \frac{3 \cdot 2 \cdot 1}{6} = 8.$$

Hence without the blank it is $8 - 1 = 7$.

Example 61. In how many ways can a different combination containing 4 flowers be made from 3 kinds of flowers if repetitions are allowed?

In the last formula 1 is subtracted each time any kind of flower is taken, and in all there are $3 - 1 = 2$ subtracted.

If repetitions are allowed it will not be subtracted, and there will be $3 - 1 = n - 1$ more kinds of flowers at the end than there would be if it were subtracted.

But if repetitions are allowed there are $n = 3$ kinds of flowers at the end, and if the 1 is successively added instead of subtracted we can use the same formula.

Hence the total number of different combinations, including a blank, is

$$1 + n + \frac{n(n+1)}{\underline{|2}} + \frac{n(n+1)(n+2)}{\underline{|3}} + \text{etc.}$$

$$= \Sigma \frac{\underline{|r + n - 1}}{\underline{|r}\,\underline{|n - 1}}.$$

$$= 1 + 3 + \frac{3 \cdot 4}{\underline{|2}} + \frac{3 \cdot 4 \cdot 5}{\underline{|3}} + \frac{3 \cdot 4 \cdot 5 \cdot 6}{\underline{|4}} + \text{etc.}$$

And the number of different combinations each containing 4 flowers is $\frac{3 \times 4 \times 5 \times 6}{\underline{|4}} = 15$.

Example 62. How many dominoes are there in a set made from the seven numbers from 0 to 6 if doubles are allowed.

There are $\frac{7 \cdot 8}{\underline{|2}} = 28$.

Example 63. How many dominoes are there in a set made from the 8 numbers from 0 to 7?

There are $\frac{8 \cdot 9}{\underline{|2}} = 36$.

Example 64. In how many ways can a bouquet of 10 flowers be made from 4 kinds of flowers ?

$$\text{In } \frac{\underline{|10 + 4 - 1}}{\underline{|10}\ \underline{|3}} = \frac{\underline{|13}}{\underline{|10}\ \underline{|3}} = 286 \text{ ways.}$$

Example 65. In how many ways can 5 coins be selected from 3 kinds of coins ?

$$\text{In } \frac{\underline{|5 + 3 - 1}}{\underline{|5}\ \underline{|3 - 1}} = \frac{\underline{|7}}{\underline{|5}\ \underline{|2}} = 21 \text{ ways.}$$

Proposition 3.

If the chance that one thing will happen is p_1, and after this has happened the chance that another thing will happen is p_2, and after these have happened the chance that another thing will happen is p_3, and so on, the chance that they will all happen in this order is $p_1 p_2 p_3$, and so on.

Say the first thing has a_1 chances to happen and a_2 chances to fail, and after this has happened the second thing has b_1 chances to happen and b_2 chances to fail, and after these have happened the third thing has c_1 chances to happen and c_2 chances to fail, and so on.

Hence $p_1 = \frac{a_1}{a_1 + a_2}$, $p_2 = \frac{b_1}{b_1 + b_2}$, $p_3 = \frac{c_1}{c_1 + c_2}$, and so on, by Definition 3.

Hence in this order all the things have $a_1 b_1 c_1$, and so on, chances to happen and $(a_1 + a_2)(b_1 + b_2)(c_1 + c_2)$, and so on, chances to happen and fail, by Definition 2 and Axiom 8.

Hence the chance that all will happen in this order is

$$\frac{a_1\, b_1\, c_1 \text{ etc.}}{(a_1 + a_2)\ (b_1 + b_2)\ (c_1 + c_2) \text{ etc.}} = p_1\ p_2\ p_3 \text{ etc., by}$$
Definition 3.

Hence the Proposition is true.

Note.—This Proposition follows directly from Definition 3 and Axiom 8.

Example 66. If there is the same number of boys and girls in any country, what is the chance that a family of 7 children will be all boys ?

The chance that any child is a boy is $p = \frac{1}{2}$.

Hence the chance that 7 children are all boys is

$p^7 = (\frac{1}{2})^7 = \frac{1}{128}$, by Proposition 3 and Definition 6.

Example 67. If the chance to win is p, what is the chance that a person will fail every time for n trials ?

The chance that he will fail every time is q^n, where $q = 1 - p$, by Definitions 3 and 6 and Axiom 8.

If he is throwing one die in the game of Dice, the chance of throwing ace is $\frac{1}{6}$, and the chance to fail is $\frac{5}{6}$, since the die has 6 sides, one of which is ace and 5 are not ace.

If $n = 10$ the chance to fail every time for 10 trials is $(\frac{5}{6})^{10} = \cdot 1615$.

And the chance that he will not fail every time, that is, that he will get ace at least once, is $1 - q^{10} = \cdot 8385$.

If he is throwing 3 dice each time instead of one the chance that he will fail to get an ace in one throw is

$\left(\frac{5}{6}\right)^3 = \left(\frac{11-1}{11+1}\right)^3 = \frac{11-3}{11+3} = \frac{4}{7}$, by Example 28.

And the chance that he will fail every time to get an ace for n throws is $(\frac{5}{6})^{3n} = (\frac{4}{7})^n$, by Definition 14.

If $n = 10$ the chance that he will fail every time for 10 throws is $(\frac{5}{6})^{30} = \cdot 0042$.

Example 68. Hence the chance that he will not fail every time is $1 - \cdot 0042 = \cdot 9958$. That is, the chance that he will get an ace at least once in 10 throws is ·9958.

Example 69. With one die, how many throws must he make before it is a chance of ·01, or 1 in 100, that he will fail every time to get an ace ?

$$q^n = (\tfrac{5}{6})^n = \cdot 01.$$

Hence $n = 25{\cdot}3$.

Hence the chance is $1 - \cdot 01 = \cdot 99$ that he will not fail every time for 25·3 throws, by Definition 3.

Example 70. With 3 dice, how many throws must he make before it is a chance of ·01 that he will fail every time to get an ace ?

$$(\tfrac{5}{6})^{3n} = \cdot 01.$$

Hence $n = 8{\cdot}4$.

Example 71. Thus it is a chance of $1 - \cdot 01 = \cdot 99$, or 99 to 1, that he will get at least one ace with 3 dice in 8·4 throws.

Example 72. If in Example 70 every time were put after ace instead of after fail, what is the answer ?

The chance that he will get an ace every time is

$$p^{3n} = (\tfrac{1}{6})^{3n}.$$

Hence the chance that he will not get an ace every time is $1 - (\frac{1}{6})^{3n} = \cdot 01$.

Hence $n = \cdot 002$.

Example 73. If r is the number of dice and s the

number of sides to each die, and y the chance of winning every time

$$y = p^{rn} = \left(\frac{1}{s}\right)^{rn}.$$

Example 74. And the chance to fail every time is $(1 - p)^{rn} = q^{rn} = \left(\frac{s-1}{s}\right)^{rn}$.

Example 75. The chance that he will not win every time is $1 - \left(\frac{1}{s}\right)^{rn}$.

Example 76. The chance that he will not fail every time is $1 - \left(\frac{s-1}{s}\right)^{rn}$.

Example 77. The chance of at least one failure is $1 - p^{rn} = 1 - \left(\frac{1}{s}\right)^{rn}$.

Example 78. The chance of at least one win is $1 - q^{rn} = 1 - \left(\frac{s-1}{s}\right)^{rn}$.

Example 79. If in a country the chance of blindness is ·001, and the chance of deafness is ·01, and the chance of dumbness is ·1, what is the chance that a person will have all three ailments?

It is assumed that the ailments are independent of one another, that is, the fact of a person having one or more of the ailments does not render him more or less subject to having one or more of the others. The fact that when they are not independent they cause irregularities in chance is used by the authorities to find how much they depend on one another.

The chance is $p_1 p_2 p_3 = \cdot 001 \times \cdot 01 \times \cdot 1 = \cdot 000001$.

Example 80. What is the chance that a person will not have all the ailments ?

It is $1 - p_1 p_2 p_3 = \cdot 999999$.

Example 81. What is the chance that a person will have none of the ailments ?

It is $q_1 q_2 q_3 = \cdot 999 \times \cdot 99 \times \cdot 9 = \cdot 890109$, by Proposition 3 and Definition 6.

Example 82. What is the chance that a person will have some of the ailments ?

It is $1 - q_1 q_2 q_3 = \cdot 109891$.

Example 83. What is the chance that a person will be both blind and deaf ?

It is $p_1 p_2 = \cdot 001 \times \cdot 01 = \cdot 00001$.

Example 84. What is the chance that a person will not be both blind and deaf ?

It is $1 - p_1 p_2 = 1 - \cdot 00001 = \cdot 99999$.

Example 85. What is the chance that a person will be both blind and deaf and not dumb ?

It is $p_1 p_2 q_0 = \cdot 001 \times \cdot 01 \times \cdot 9 = 000009$, by Definition 3 and Axiom 8.

Example 86. What is the chance that a person will not be both blind and deaf and not dumb ?

It is $1 - p_1 p_2 q_3 = \cdot 999991$.

Example 87. What is the chance that a person will be neither blind nor deaf and will be dumb ?

It is $q_1 q_2 p_3 = \cdot 999 \times \cdot 99 \times \cdot 1 = \cdot 098901$.

Example 88. What is the other complementary chance ?

It is $1 - q_1 q_2 p_3 = \cdot 901099$.

Proposition 4.

If out of any number of things one must happen, and only one can happen, and the chance that one will happen is p_1 and the chance that another will happen is p_2, and the chance that another will happen is p_3, and so on, the chance that either the first or second will happen is $p_1 + p_2$, and the chance that one of any two or more will happen is the sum of the chances of the two or more, and the chance that some one will happen is 1.

The chance that the first will happen is p_1 of all the chances, and the chance that the second will happen is p_2 of all the chances, by Definition 3.

Hence the chance that the first or second will happen is $p_1 + p_2$ of all the chances.

Hence the chance that the first or second will happen is $p_1 + p_2$, by Definition 3.

And similarly for one of any two or more.

And since some one must happen the sum of all the chances favourable to some one happening is equal to the sum of all the chances.

Hence the chance that some one will happen is 1, by Definition 3.

Hence the Proposition is true.

Note.—This Proposition follows directly from Definition 3.

Example 89. If there is the same number of boys and girls in a country, what is the chance that a family of 5 children will be either all boys or girls ?

It is $(\frac{1}{2})^5 + (\frac{1}{2})^5 = 2(\frac{1}{2})^5 = \frac{1}{16}$, by Definition 3 and 6 and Axiom 8.

Example 90. If the chance that an unknown thing

is a boy is $\frac{1}{2}$ and the chance that it is a girl is $\frac{1}{2}$, what is the chance that it is either a boy or girl?

It is $\frac{1}{2} + \frac{1}{2} = 1$, by Definition 3.

Example 91. If the chance that A will win is $\frac{1}{3}$ and the chance that B will win is $\frac{1}{6}$ and the chance that C will win is $\frac{1}{10}$, what is the chance that A, B or C will win?

It is $\frac{1}{3} + \frac{1}{6} + \frac{1}{10} = \frac{3}{5} = \cdot 6$, by Proposition 4.

PROPOSITION 5.

If the chance that a thing will happen in any trial is p and the chance that it will fail is q, the chance that it will happen $n - r$ times and fail r times in n trials is

$\frac{\underline{|n}}{\underline{|n-r}\,\underline{|r}}\, p^{n-r} q^{r}$, and the sum of all the chances is

$$(p+q)^n = p^n + \frac{\underline{|n}}{\underline{|n-1}\,\underline{|1}}\, p^{n-1} q^1 +$$

$$\frac{\underline{|n}}{\underline{|n-2}\,\underline{|2}}\, p^{n-2} q^2 + \frac{\underline{|n}}{\underline{|n-3}\,\underline{|3}}\, p^{n-3} q^3 + \text{and so on,}$$

$$= p^n + np^{n-1} q + \frac{n(n-1)}{\underline{|2}} p^{n-2} q^2 +$$

$$\frac{n(n-1)(n-2)}{\underline{|3}} p^{n-3} q^3 + \text{and so on} = 1.$$

Wins can be arranged in only one order, since they are all regarded as alike, by Definition 6.

Hence the chance that it will happen $n - r$ given times is p^{n-r}, by Proposition 3.

Similarly the chance that it will fail r given times is q_r, by Proposition 3.

Hence the chance that it will happen $n - r$ given times and fail r given times is $p^{n-r} q^r$, by Proposition 3.

And n different things can be divided up into $n - r$ wins and r failures in $\frac{\lfloor n}{\lfloor n-r \, \lfloor r}$ ways, by Proposition 1, since each thing is composed of a single element.

Hence the chance of $n - r$ wins and r failures in a series of n trials is

$$p^{n-r} q^r \times \frac{\lfloor n}{\lfloor n-r \, \lfloor r} = \frac{\lfloor n}{\lfloor n-r \, \lfloor r} p^{n-r} q^r,$$

by Proposition 4.

And r can successively take all the integral values from 0 to n.

Hence by successively giving to r the values 0, 1, 2, 3, and so on, and adding the results we will get the sum of all the chances, by Proposition 4.

And since p of all the chances is favourable and q of all the chances is unfavourable $p + q = 1$, by Definition 3.

Hence the Proposition is true.

Example 92. At the game of Head or Tail, what is the chance of 4 heads and 2 tails in 6 trials ?

$$p = q = \tfrac{1}{2}.$$

$$y = \frac{\lfloor 6}{\lfloor 4 \, \lfloor 2} (\tfrac{1}{2})^4 (\tfrac{1}{2})^2 = \frac{15}{64}.$$

Example 93. What is the chance that there will be less than 4 tails in 6 trials ?

The chance of $n - r$ heads and r tails in n trials is

$$y = \frac{\lfloor n}{\lfloor n-r \, \lfloor r} p^{n-r} q^r.$$

Give to r all the integral values from 0 to 3 in this and add the results.

$$\text{Hence } y = \frac{\underline{|6}}{\underline{|6-0}\ \underline{|0}}(\tfrac{1}{2})^6 (\tfrac{1}{2})^0 + \frac{\underline{|6}}{\underline{|6-1}\ \underline{|1}}(\tfrac{1}{2})^5 (\tfrac{1}{2})^1 +$$

$$\frac{\underline{|6}}{\underline{|4}\ \underline{|2}}(\tfrac{1}{2})^4 (\tfrac{1}{2})^2 + \frac{\underline{|6}}{\underline{|3}\ \underline{|3}}(\tfrac{1}{2})^3 (\tfrac{1}{2})^3.$$

$$= \frac{1}{64} + \frac{6}{64} + \frac{15}{64} + \frac{20}{64} = \frac{42}{64} = \cdot 6563.$$

Example 94. What is the chance that the error will be less than 1 ?

It is the chance of no error or

$$\frac{\underline{|6}}{\underline{|3}\ \underline{|3}}(\tfrac{1}{2})^3 (\tfrac{1}{2})^3 = \frac{20}{64} = \cdot 31,$$

by Definitions 11 and 14.

Example 95. What is the chance of no error ?

It is $\frac{\underline{|n}}{\underline{|np}\ \underline{|nq}} p^{np} q^{nq}$, by Definition 3 and 11.

Example 96. What is the chance of an error of $+x = +3$ at the end of 18 trials if $p = \frac{1}{6}$ and $q = \frac{5}{6}$?

$$\text{The chance is } y = \frac{\underline{|n}}{\underline{|np+3}\ \underline{|nq-3}} p^{np+3} q^{nq-3}.$$

$$= \frac{\underline{|18}}{\underline{|6}\ \underline{|12}}\left(\frac{1}{6}\right)^6 \left(\frac{5}{6}\right)^{12},$$

by Definition 11 and 14.

Example 97. If A throws a die 6 times and a coin 6 times, what is the chance that he will get either 4 aces or 4 heads ?

A die has 6 sides, one of which is ace and 5 are not ace.

Hence $p = \frac{1}{6}$ and $q = \frac{5}{6}$.

A coin has 2 sides, of which one is head and one is not head.

Hence $p = q = \frac{1}{2}$.

We must add the two chances $\frac{125}{15552}$ and $\frac{15}{64}$ = ·0080 + ·2344 = ·2424, by Proposition 4.

Example 98. What is the chance of both 4 aces and 4 heads?

It is the product of the two chances $\frac{125}{15552}$ and $\frac{15}{64}$, by Definition 3.

Example 99. What is the chance of both 4 aces and less than 4 heads?

It is the product of the two chances $\frac{125}{15552}$ and $\frac{42}{64}$, by Definition 3.

Example 100. If the bookmakers at a race put up the following odds against the following horses winning, has the public a fair chance?

Mary: 1 to 3;
Charley: 1 to 2;
Jack: 2 to 3;
Katy: 3 to 1;
Eva: 99 to 1.

If the same amount of money, say $1,000, is involved about each horse the public pay in

$\frac{3}{4}$	or	·75	of $1,000 =	$750	on Mary;
$\frac{2}{3}$	or	·677	of $1,000 =	$667	on Charley;
$\frac{3}{5}$	or	·6	of $1,000 =	$600	on Jack;
$\frac{1}{4}$	or	·25	of $1,000 =	$250	on Katy;
$\frac{1}{100}$	or	·01	of $1,000 =	$10	on Eva.
	or	2·277	of $1,000 =	$2,277	on all.

Now no matter which horse wins the bookmakers will not have to pay out as much as **$1,000**, because this is all the money involved about any horse.

The sum of the fractions should be 1 instead of 2·277, by Proposition 4.

If it is less than 1 any player can bet on all the horses and certainly win, if the bookmakers pay up.

Example 101. If the above odds were

Mary : 1 to 1 ;
Charley : 4 to 1 ;
Jack : 9 to 1 ;
Katy : 19 to 1 ;
Eva : 99 to 1 ;

a player could bet

$50 on Mary ;
$20 on Charley ;
$10 on Jack ;
$5 on Katy ;
$1 on Eva.

and he would get $100, no matter which horse won, having paid in only $86.

The public might give the bookmakers something over 10 per cent. for their trouble and risk, and let them make the sum of the fractions 1·10 instead of 1, seeing that they cannot always get the same amount of money involved about each horse.

Example 102. The slate might fairly be

Mary : 2 to 3 ;
Charley : 5 to 3 ;
Jack : 9 to 1 ;
Katy : 19 to 1 ;
Eva : 99 to 1.

Here the public pay in on

Mary:	\$600;
Charley:	\$375;
Jack:	\$100;
Katy:	\$50;
Eva:	\$10.

making \$1,135 in all.

And no matter which horse wins they cannot get back as much as \$1,000.

The chance that some horse will win is 1. Hence the sum of the chances of all the horses winning is 1, since only one horse can win, by Proposition 4.

But the bookmakers sometimes make it 1·15, 1·50, 2·50, 6·00, and even as high as 10 and 20. The odds advertised in the papers sometimes are as bad as this.

If bookmakers are expert in getting the same amount of money involved about each horse it is a safe game for them. They constantly change the odds for this purpose. But they often gamble on particular horses and sometimes go broke.

PROPOSITION 6.

If we know nothing of anything of constant habits except that it has happened m times and failed n times, the chance that it will happen next time is $\frac{m+1}{m+n+2}$.

It has happened m times and failed n times.

And the next time it must either happen or fail.

Hence at the next time there are $m+n+2$ possible events, of which $m+1$ are wins and $n+1$ are failures, by Definition 15. We are learning its habits.

And, since we know nothing more of the thing, all these possible events seem equally likely, by Axiom 10.

Hence the chance that the next event will be a win is $\frac{m+1}{m+n+2}$, by Definition 3.

Hence the Proposition is true.

Example 103. If a thing has happened 6 times and failed 3 times, and we have no other knowledge about it, what is the chance that it will happen next time?

It is $\frac{m+1}{m+n+2} = \frac{7}{11}$.

Example 104. What is the chance that it will fail next time?

It is $\frac{n+1}{m+n+2} = \frac{4}{11}$.

Or it is $1 - \frac{7}{11} = \frac{4}{11}$, by Definition 3.

Example 105. If the first $m+n$ events are m wins and n failures, what is the chance that the next $r+s$ events will be r wins and s failures, if we have no other knowledge?

The number of ways in which there can be r more wins is the same as the number of ways in which a different combination of m wins can be made from $m+r$ wins, since its habits do not change, and when m are taken r are left.

And the number of ways in which there can be s more failures is the same as the number of ways in which a different combination of n failures can be made from $n+s$ failures.

Hence the number of ways in which there can be both r more wins and s more failures is

$$\frac{\lfloor m+r}{\lfloor m \, \lfloor r} \times \frac{\lfloor n+s}{\lfloor n \, \lfloor s},$$

by Proposition 2 and Axiom 8.

At the next event there are $m + n + 2$ possible events, by Proposition 6.

Hence if there are $r + s$ more events there will be $m + n + 2 + (r + s - 1) = m + n + 1 + r + s$ possible events altogether, since the next event must be either a win or a failure.

And a different combination of $r + s$ events can be made from $m + n + 1 + r + s$ possible events in $\dfrac{(m + n + 1 + r + s)!}{(m + n + 1)!\,(r + s)!}$ ways, by Proposition 2.

Hence the chance that the $r + s$ events will be r wins and s failures is

$$\frac{\dfrac{(m+r)!}{m!\,r!} \times \dfrac{(n+s)!}{n!\,s!}}{\dfrac{(m+n+1+r+s)!}{(m+n+1)!\,(r+s)!}}, \text{ by Definition 3.}$$

Example 106. If we saw a soldier hit a target, or make an inner, or an outer, or do anything else, 5 times, and fail 3 times, and we had no other knowledge, what is the chance that he will win twice and fail once in 3 more trials?

It is $\frac{21}{55}$.

Example 107. If we saw him win 5 times and fail 3 times the chance that he will win next time is ·6.

Example 108. If we saw him hit twice and miss twice the chance that he will hit next time is $\frac{1}{2}$ = ·5. p and q change at each trial.

If we watch a great many shots p and q will change very little after $m + n$ becomes large.

A man may have very bad luck in his first few shots, and runs of bad luck at any time. But after a great many trials the *a priori* and *a posteriori* chances can hardly differ much since the number of chances of error is only proportional to $\sqrt{(m + n + 1)\,pq}$, by Axiom 3 and Proposition 3.

The *a priori* and *a posteriori* methods can be compared and used to check each other in any case where we can assume *a priori* chances.

The difference between the *a posteriori* and *a priori* chances, that is, the changing of p and q as our knowledge is increased, is the source of many strange traditions.

If a man follows a very limited experience his *a posteriori* chance may be very different from the true *a priori* chance.

When a scientist tries to prove anything, if he cannot get the *a priori* chance, he makes a very large number of observations in order that his *a posteriori* chance will differ but little from the true *a priori* chance. If we read any great book on modern experimental science the almost infinite number of experiments, seemingly needless, that are made is sometimes tedious. And these observations are made by chance.

The observations made by Spiritualists are not always by chance, and the unfavourable results are not always counted.

If the desired result happens m times and fails n times, they sometimes have a tendency to take $\frac{m+1}{m+2}$ instead of $\frac{m+1}{m+n+2}$ as the chance that it will happen next time.

Many people, like the Pacifists, have a dislike for checking their *a priori* results with the *a posteriori* results

There are different opinions on all subjects because each person sees a different part. The more parts that different people see the less difference there is in their opinions, and the more nearly will the *a posteriori* results agree with the true *a priori* results.

When evidence is taken from only one part it is said to be *ex parte* evidence.

Proposition 7.

The chance that the error at the end of any series will be greater than $x = sr$, where r is the Median error at the end of the series, is $1 - P = \dfrac{(\frac{1}{2})^{\frac{s+1}{2}}}{\lfloor s}$, very nearly, if s is not much less than 1, where P is the chance that the error will be less than $x = sr$.

Any error is a combination of errors all of the same contradictory, for things which contradict each other cannot be added together, because they cannot exist at the same time, by Definition 8 and Axiom 3.

The Median error is the error that is as likely to be exceeded as not. We will denote it by r.

Let there be any $2s$ errors all belonging to the same contradictory.

The chance that the first will be less than r is $\frac{1}{2}$, and the chance that it will be greater than r is $\frac{1}{2}$, by Definition 3.

And the chance that the second will be less than r is $\frac{1}{2}$, and the chance that it will be greater than r is $\frac{1}{2}$.

Hence the chance that each will be less than r is

$\frac{1}{2} \times \frac{1}{2}$, and the chance that each will be greater than r is $\frac{1}{2} \times \frac{1}{2}$, by Proposition 3 and Definition 6.

Hence the chance that one will be less and the other greater than r is $\frac{1}{2} \times \frac{1}{2} \times 2 = \frac{1}{2}$, by Axiom 9.

Similarly the chance that the third and fourth will be one less and the other greater than r is $\frac{1}{2}$.

Hence the chance that the first four errors will be two less and two greater than r is $\frac{1}{2} \times \frac{1}{2} = (\frac{1}{2})^2$, by Proposition 3 and Definition 6.

Similarly the chance that the $2s$ errors will be s less and s greater than r is $(\frac{1}{2})^s$.

Any of the errors is equally likely to be less or greater than r.

And the chance that s will go on one side and s on the other is $(\frac{1}{2})^s$.

Now all we have to consider is the different positions that the errors can occupy after they are put half on one side and half on the other, so as to arrange them in the order in which they occurred, by Proposition 3. And all orders are equally likely.

The s errors each less than r can be arranged in $\underline{|s}$ ways, and the s errors each greater than r can be arranged in $\underline{|s}$ ways, by Axiom 9.

And these two groups can be arranged in $\underline{|2}$ ways, by Definition 5 and Axiom 9.

Hence the $2s$ errors can be arranged into two groups of s errors each less and s errors each greater than r, and these two groups arranged, in $2\,\underline{|s}\,\underline{|s}$ orders, by Axiom 8.

Hence the chance that the $2s$ errors will occur in a given order with s errors each less and s errors each greater than r is $\dfrac{(\frac{1}{2})^s}{2\,\underline{|s}\,\underline{|s}}$.

The chance that the s errors each less than r will all occur in this order is the same as the chance that the s errors each greater than r will all occur in this order, for the chance of an error less than r is the same as the chance of an error greater than r.

Hence the chance that the the s errors each greater than r will all occur in this order is $\sqrt{\dfrac{(\frac{1}{2})^s}{2\,\lfloor\underline{s}\,\lfloor\underline{s}}} = \dfrac{(\frac{1}{2})^{\frac{s+1}{2}}}{\lfloor\underline{s}}$, by Proposition 3.

And the combination of the s errors each greater than r will give a total error greater than sr, since they must all belong to the same contradictory.

And the chance that each of these errors will be greater than r must be very nearly the same as the chance that their sum will be greater than sr. This can be easily seen if they are arranged in order of magnitude.

But if s is much less than 1 we have no group of errors each greater than r, and the formula will not be precise.

Hence the Proposition is true.

Example 109. Proposition 7 is correct to not less than about three decimal places for all values of s not less than 1.

For values of s less than 1 an approximate result is $P = \dfrac{s}{2}$.

Since the factorial of any fraction less than 1 is not far from 1 it may be taken as 1 for rough work.

$\lfloor\underline{\cdot 05} = \cdot 97$, $\lfloor\underline{\cdot 1} = \cdot 95$, $\lfloor\underline{\cdot 2} = \cdot 92$, $\lfloor\underline{\cdot 3} = \cdot 90$,

$\lfloor\underline{\cdot 4} = \cdot 89$, $\lfloor\underline{\cdot 5} = \cdot 89$, $\lfloor\underline{\cdot 6} = \cdot 89$, $\lfloor\underline{\cdot 7} = \cdot 91$,

$\lfloor\underline{8\cdot} = \cdot 93$, $\lfloor\underline{\cdot 9} = \cdot 96$, $\lfloor\underline{\cdot 95} = \cdot 98$, to two decimal places.

In any case where s is fractional $\underline{|s}$ can easily be found from these values.

Thus $\underline{|4{\cdot}4} = 4{\cdot}4 \times 3{\cdot}4 \times 2{\cdot}4 \times 1{\cdot}4 \times \underline{|{\cdot}4}$.

If the factorial of any fraction less than 1 is taken as 1 the result may be out about 12 per cent. as a maximum, and the correction can easily be made for approximate results.

Example 110. What is the chance that an error will be greater than $3r$?

It is $\dfrac{(\frac{1}{2})^{\frac{3+1}{2}}}{\underline{|s}} = \dfrac{(\frac{1}{2})^2}{\underline{|3}} = \dfrac{1}{24} = {\cdot}042$, very nearly.

Proposition 8.

Contradictory lines are at right angles to each other.

Contradictory lines tend to contradictories, by Definition 9.

If any two lines are not at right angles to each other either line has a component in the direction of the other, by Axiom 6.

That is, they tend to things that have a common element.

But contradictories cannot have a common element, by Axiom 1.

Hence the Proposition is true.

Proposition 9.

Lines tending to the real and the imaginary are at right angles to each other.

If two lines are not at right angles to each other either has a component in the direction of the other, by Axiom 6.

That is, they tend to things that have a common element.

But the real and the imaginary cannot have a common element, by Axiom 2.

Hence the Proposition is true.

Proposition 10.

All lines tending to contradictories are in the same plane, called the real plane.

Two contradictories are an imaginary pair, by Definitions 8 and 16. Hence any line tending to both is imaginary, by Definition 17.

Hence any line tending to either contradictory must be at right angles to an imaginary line, by Propositions 8 and 9.

Hence all lines tending to contradictories must be in the same plane.

And any line tending to either is real.

Hence the Proposition is true.

Proposition 11

If the elements are all very small and p is the chance to win and q the chance to fail in any game, the sum of the chances of a pair of contradictories in a series of n games is $2\pi\, npq = 2\pi\, x_0^2 = \pi\, r_0^2$, and the sum of the chances of error in the series is

$\sqrt{2\pi npq} = \sqrt{2\pi\, x_0^2} = \sqrt{\pi\, r_0^2}$, where x_0 is the Normal error in any dimension of space, and r_0 is the Normal error in a plane target.

And the error sometime during a series of games is the same as the error in a plane target.

And the sum of the chances of n errors one in each dimension of space is $(\sqrt{2\pi\, npq})^n$.

And the Normal error x_o in any dimension of space is $x_a \sqrt{\frac{\pi}{2}}$, where x_a is the average error in any dimension.

And the average error in a plane target is $r_a = \frac{\pi}{2} x_a$.

And the average error in a sphere is $R_a = r_a \sqrt{\frac{\pi}{2}}$.

And the normal error in a plane target is $r_o = \sqrt{2x_0^2}$.

And the normal error in a sphere is $R_o = \sqrt{3x_0^2}$.

In n games there will be np wins and nq failures if no wins and failures tend to occur at the same time, because p of all the games will be wins and q of all the games will be failures, by Definition 3.

Error is caused by yes and no tending to occur at the same time, by Definition 8 and Axiom 3.

The chance that a win and a failure will tend to occur at the same time is pq, by Proposition 3.

Hence pq is the average number of pairs of contradictories for one game, by Definitions 3 and 8.

Hence npq is the average number of pairs of contradictories for n games, by Definition 2 and 8.

And npq pairs of contradictories are formed from $\sqrt{npq}$ wins and $\sqrt{npq}$ failures, since the wins tend to occur at the same time as the failures, by Axioms 3 and 8.

Hence $\sqrt{npq}$ is the normal number of wins or failures in error at the end of the series, but not the average number, by Definition 12.

Hence $\sqrt{npq} = x_0$ is the Normal error at the end of the series, by Definition 12.

But since wins and failures at the same time are contradictories $\sqrt{npq}$ wins are at right angles to $\sqrt{npq}$ failures, by Axiom 4 and Proposition 8.

And they may take any direction in a plane so long as they are at right angles to each other, by Proposition 10.

Hence their resultant $\sqrt{2npq}$ may take any direction in a plane, by Axiom 7.

Hence the sum of the chances of a pair of contradictories during the series is the area swept over by the radius $\sqrt{2npq} = r_0$ as it revolves around once, or $2\pi npq = 2\pi x_0{}^2 = \pi r_0{}^2$.

And r_0 is the normal error during the series, by Definition 12.

Hence the sum of the chances of wins or failures in error during the series is $\sqrt{2\pi npq} = \sqrt{2\pi x_0{}^2} = \sqrt{\pi r_0{}^2}$, by Axioms 3 and 8.

And the sum of the chances of n errors, one in each dimension of space, is $(\sqrt{2\pi\, npq})^n$, by Definition 12 and Axioms 5 and 8, since in space three straight lines can be drawn at right angles to each other, and each of these lines is called a dimension of space.

An error in a line is in one dimension, an error in a plane is in two dimensions and an error in a sphere is in three dimensions of space.

Again a vector is anything that has both magnitude and direction.

Hence errors are vectors. And they can take any direction in a plane, by Proposition 10.

Hence the sum of the chances of error is proportional to the amount of its possible change in direction, by Definition 2.

It is equal to the error multiplied by the circumference of a circle over the diameter, since the error may be either positive or negative, by Definition 18.

The error r_1 in a line can take any direction in a plane, by Proposition 10.

Now the circumference of a circle includes all the possible directions in a plane, and none of them more than once, and half of them are in opposite directions to the other half, by Axiom 10.

But positive and negative errors cannot form parts of the same error, by Definition 8.

Hence the sum of the chances of error in a line is $\frac{\Sigma \pi r_1}{n} = \pi x_a$, or half the circumference of a circle of radius x_a.

It is x_a multiplied by the angle π through which it can turn, by Definition 18.

It is equal to π average errors in a line.

Hence $\pi x_a = \sqrt{2\pi x_0{}^2} = \sqrt{2\pi npq}$.

Hence $x_a = x_0 \sqrt{\frac{2}{\pi}} = \sqrt{\frac{2npq}{\pi}} = \sqrt{\frac{r_0{}^2}{\pi}}$.

Again any error r_1 in a line is measured up or down or to the right or to the left in a plane target.

Now if each of these four errors is turned through a right angle they pass over the whole target, and include all the directions of the radii, which are the errors in a plane target.

Hence the error in a plane target is equal to the error

in a line multiplied by a right angle, or $\frac{\pi}{2}$, by Definition 18, since the errors are all in the same plane, that of the target.

Hence $\frac{\Sigma r_2}{n} = \frac{\pi}{2} \cdot \frac{\Sigma r_1}{n}$. where r_2 is the error in two dimensions.

Hence $r_a = \frac{\pi}{2} x_a$.

And any error r_2 in a plane is measured parallel to the plane target or to a plane at right angles to the plane target in a sphere.

And the pairs of contradictories are at right angles to the planes containing the errors, by Definition 16 and Proposition 9.

And if these two planes at right angles to each other are both turned through a right angle about their common diameter they pass through a sphere.

And as they are turned through a right angle the pairs of contradictories at right angles to the planes also turn through a right angle

Hence the average error in a sphere is

$R_a = \sqrt{\left(\frac{\Sigma r_3}{n}\right)^2} = \sqrt{\frac{\pi}{2}\left(\frac{\Sigma r_2}{n}\right)^2} = r_a \sqrt{\frac{\pi}{2}}$, where r_3 is the error in three dimensions of space, by Definition 2.

And the normal error in a plane target is

$r_0 = \sqrt{\frac{\Sigma r_2^2}{n}} = \sqrt{2x_0^2}$, by Definition 12 and Axioms 3 and 7.

And the normal error in a sphere is $R_0 = \sqrt{\frac{\Sigma r_3^2}{n}} = \sqrt{3x_0^2}$ by Definition 12 and Axioms 3 and 7.

Hence the Proposition is true.

P.S.—The actions of contradictories have no effect on each other and must be taken separately, by Axiom 5. x wins carry a point from 0 to x in one dimension and x failures take it on a distance of x in another dimension of space, and the total effect, or error, is the resultant which is the line joining the beginning with the end, by Axiom 3, and this resultant must be either all wins or all failures, by Definition 8, and can take any direction in the plane containing the two contradictories, by Proposition 10, and it is equally likely to be all wins or all failures, by Axiom 10.

And wins and failures are equally likely in any dimension, by Axioms 3 and 10.

And for successive errors we must add the chances of a pair of contradictories, because each error has a different set of contradictories, by Definition 8, and the number of chances of a pair of contradictories is normally proportional to the number of errors, by Definition 12.

Hence for n successive errors in a line the sum of the chances of error is $\sqrt{n\,(\pi\, x_a)^2} = \pi x_a \sqrt{n}.$

And in a plane target is $\pi r_a \sqrt{n},$ since there is the same number of errors.

And in a sphere is $\pi \mathrm{R}_a \sqrt{n}.$

Hence we must always add the squares of successive errors, by Axiom 3.

And in finding the average error we must disregard the signs, by Axiom 3 and Definition 8.

Integration. If c is a very small element of x they usually write dx for c, which means that dx is a very small change or differential of x as x changes continuously in value, and $(x + c)^n = x^n + nx^{n-1}c,$ very nearly, by Proposition 5, and exactly when c gradually becomes 0.

Thus when c is added to x, $nx^{n-1}c$ is added to x^n.

Hence when c is a very small element of x, $nx^{n-1}c$ is a very small element of x^n, by Definition 7.

And this is true no matter how x varies so long as the variation is continuous, and x^n is equal to the sum of all the possible values of $nx^{n-1}c$ from $x = 0$ to $x = x$, by Definition 7.

For this reason x^n is called the integral of $nx^{n-1}c$.

Thus when a quantity is continuously varying addition is called integration, and the sum of all the differentials $nx^{n-1}c$ for all the possible values of x from 0 to x is called the integral of $nx^{n-1}dx$ from 0 to x, and is expressed by $\int_0^x nx^{n-1}dx = x^n$.

And the differential of x^n by $d(x_n) = nx^{n-1}dx$.

When e to any power is expanded by Proposition 2 each term can be integrated separately very easily, and the sum of all the separate integrals gives the total integral.

Proposition 12.

If the elements are all very small the chance of an error of magnitude r is $\frac{2dr}{\sqrt{2\pi x_0^2}} e^{-\frac{r^2}{2x_0^2}}$ for a line or at the end of a linear series, $\frac{rdr}{x_0^2} e^{-\frac{r^2}{2x_0^2}}$ for the radius of a circle or sometime during a linear series as of games, and $\frac{4\pi r^2 dr}{(\sqrt{2\pi x_0^2})^3} e^{-\frac{r^2}{2x_0^2}}$ for the radius of a sphere, where x_0 is the Normal error in a line or linear series.

Any error in any dimension is the resultant of two equal contradictory components, only one but either, of which can occur, by Definition 8 and Proposition 11.

And the sum of the chances of a pair of contradictories for the whole of any linear series is $2\pi npq = 2\pi x_0^2 = \pi r_0^2$, by Proposition 11.

If the error is r the number of pairs of contradictories is $\frac{\pi r^2}{\pi r_0^2} = s$ times the normal number.

Now $2\pi npq$ is the sum of the chances of a pair of contradictories in any series. There may be s times this many pairs, where the change in s is continuous, that is, by very small elements, by Definition 14, and s may have any value from 0 to infinity, by Proposition 11.

And the sum of all the chances of s for all its possible values is 1, by Proposition 4.

The chance of a given combination of s things from s things, or any given combination for that matter, since all combinations from s things seem equally likely, by Axiom 10, is $\frac{1}{e^s} = e^{-s}$, by Proposition 2 and Definition 3.

$$\text{And} \int_0^\infty e^{-s}\, ds = 1.$$

Hence the chance that there will be s times the normal number of pairs of contradictories in any series is e^{-s}, by Proposition 2.

If the error is in a line the end of it may fall anywhere in $2dr$, for it may be either positive or negative in a line.

And if the error is the radius of a circle the end of it may fall anywhere in the circumference of thickness dr, that is, anywhere in $2\pi r dr$.

And if the error is the radius of a sphere its end may fall anywhere in the surface of thickness dr, that is, anywhere in $4\pi r^2 dr$.

Hence if the error is in a line the sum of the chances of an error r for all its possible values is

$\int_0^\infty e^{-s} \times 2dr = \sqrt{2\pi npq}$, by Proposition 11, since r is the resultant of two errors, only one of which, but either, can occur in one dimension of space.

And if it is the radius of a circle the sum of the chances of an error r for all its possible values is

$\int_0^\infty e^{-s} \times 2\pi r dr = 2\pi npq$, by Proposition 11, since r is the resultant of two errors one in each dimension of space.

And if it is the radius of a sphere the sum of the chances of an error r for all its possible values is

$\int_0^\infty e^{-s} \times 4\pi r^2 dr = (\sqrt{2\pi npq})^3$, by Proposition 11, since r is the resultant of three errors one in each dimension of space.

Hence the chance of an error r in a line is

$\frac{2dr}{\sqrt{2\pi npq}} e^{-\frac{r^2}{2x_0^2}}$, since the sum of the chances of r, for all its possible values must be 1, by Proposition 4.

And the chance of an error r the radius of a circle is

$$\frac{2\pi r dr}{2\pi npq} e^{-\frac{r^2}{2x_0^2}} = \frac{r dr}{2x_0^2} e^{-\frac{r^2}{2x_0^2}}.$$

And the chance of an error r the radius of a sphere is

$$\frac{4\pi r^2 dr}{(\sqrt{2\pi npq})^3} e^{-\frac{r^2}{2x_0^2}}$$

Hence the Proposition is true.

Example 111. If n is very large, what is the chance that there will be exactly n things in a combination from n things? And what is the value of $\lfloor n$?

The number of ways in which a different combination can be made from n different things is

$$e^n = 1 + n + \frac{n^2}{\lfloor 2} + \text{and so on, by Proposition 2.}$$

If there is only one thing this is

$$e = 1 + 1 + \frac{1}{\lfloor 2} + \quad \text{and so on.}$$

The first term is the tendancy to make a thing fail, and the second term is the tendancy to make it occur at the same time, since it can fail in one way and occur in one way, by Proposition 2, Definition 2 and Axiom 4.

Hence the normal error for one thing is $\sqrt{x_0^2} = \sqrt{1 \times 1} = 1$, by Proposition 11.

And the number of ways in which a pair of contradictories can be made from one thing, when all the tendencies to take it and not take it are considered, is $2\pi \times 1^2 = 2\pi$, by Proposition 11.

Hence the total number of chances of a pair of contradictories for the whole series of n things is $2\pi n$, by Proposition 11.

Hence the sum of the chances of error is $\sqrt{2\pi n}$, by Proposition 11.

If a thing is a very small element of n things, dn may be taken as 1, by Definition 14.

Hence the chance of no error is $\frac{1}{\sqrt{2\pi n}}$, by Proposition 12. But if there is no error there will be a combination of $1 \times n = n$ things at the end of the series, by Axiom 3.

Hence the chance of a combination of n things is $\frac{1}{\sqrt{2\pi n}}$.

But there can be a combination of n things in $\frac{n^n}{n!}$ ways, when the elements are all very small, by Proposition 2.

And the total number of combinations is e^n, by Proposition 2.

Hence the chance of a combination of n things is

$$\frac{\frac{n^n}{n!}}{e^n} = \frac{n^n}{n!} e^{-n}\text{, by Definition 3.}$$

Hence $\frac{1}{\sqrt{2\pi n}} = \frac{n^n}{n!} e^{-n}$, if the elements are all very small.

Hence $n! = n^n e^{-n} \sqrt{2\pi n}$, by Definition 14.

A still closer value is

$$n! = n^n e^{-n} \sqrt{2\pi n} \times \frac{24n + 1}{24n - 1}$$

When n is infinite they are both exact, because then a thing is as small an element of n things as possible.

Example 112. If the errors in 4 measurements of the same thing all made under the same set of conditions, are x_1, x_2, x_3 and x_4 respectively, what is the Normal error ?

The Normal error is such that it may be substituted for each of the errors without changing the chance that they will all occur, by Definition 12.

Hence the normal number of pairs of contradictories for the series is $\Sigma x^2 = 4x_0^2$, by Proposition 11.

Hence the Normal error is $x_0 = \sqrt{\frac{\Sigma x^2}{4}}$.

Example 113. What is the chance that the magnitude of the error will be x_5 ?

It is $\frac{2\,dx}{\sqrt{2\pi x_0^2}}\, e^{-\frac{x_5^2}{2x_0^2}}$, by Proposition 12.

Example 114. In a series of 1000 flips of a coin, what is the chance of 500 heads and 500 tails ?

$p = \frac{1}{2}$ and $q = \frac{1}{2}$, by Definition 3.

Hence if there is no error there will be $np = 1000 \times \frac{1}{2} = 500$ heads and $nq = 1000 \times \frac{1}{2} = 500$ tails in the series, by Definition 3 and Axiom 3.

Since the series is so large dn may be taken as 1, by Definition 14.

Hence the chance of no error is

$$\frac{1}{\sqrt{2\pi npq}} = \frac{1}{\sqrt{2\pi \times 1000 \times \frac{1}{2} \times \frac{1}{2}}} = \cdot 025,$$

very nearly, or once in 40 series.

Example 115. If 1 person out of 10 in the country has an automobile, what is the chance that in a district of 10,000 population there will be exactly 1000 automobiles?

$$p = \tfrac{1}{10} \text{ and } q = \tfrac{9}{10}\cdot$$

If there are exactly 1,000 automobiles there will be no error, by Definition 11.

Hence the chance is

$$\frac{1}{\sqrt{2\pi npq}} = \frac{1}{\sqrt{2\pi \times 1000 \times \frac{1}{10} \times \frac{9}{10}}} = \cdot 013,$$

very nearly.

Example 116. If 1 out of 100 men in any country differ from 67 inches in height by less than 1 inch, what is the chance that in a district of 100,000 men there will be exactly 1,000 between 66 and 68 inches in height?

$$p = \cdot 01 \text{ and } q = \cdot 99.$$

If there is no error there will be exactly 1,000 men of this height.

Hence the chance is

$$\frac{1}{\sqrt{2\pi npq}} = \frac{1}{\sqrt{2\pi \times 100{,}000 \times \cdot 01 \times \cdot 99}} = \cdot 013.$$

Proposition 13.

If all the elements are very small and in any n series all made under the same set of conditions the errors are x_1, x_2, x_3, and so on, the chance of an error greater than x sometime during any one of the series is $e^{-\frac{x^2}{2\frac{\Sigma x^2}{n}}} = e^{-t^2}$, and the chance that the error at the end of any one series will be less than x is

$$P = \frac{2}{\sqrt{2\pi\frac{\Sigma x^2}{n}}}\int_0^x e^{-\frac{x^2}{2\frac{\Sigma x^2}{n}}}dx = \frac{2}{\sqrt{\pi}}\int_0^t e^{-t^2}dt,$$

where $t = \frac{x}{\sqrt{2\frac{\Sigma x^2}{n}}}$.

By Proposition 12 the chance that an error will have the magnitude x some time during any series is

$$\frac{xdx}{x_0^2}e^{-\frac{x^2}{2x_0^2}}.$$

Hence the chance that the error will be greater than x sometime during the series is

$$\int_x^\infty \frac{xdx}{x_0^2}e^{-\frac{x^2}{2x_0^2}} = e^{-\frac{x^2}{2x_0^2}} = e^{-t^2}.$$

And by Proposition 12 the chance that the magnitude of the error at the end of the series will be x is

$$\frac{2dx}{\sqrt{2\pi x_0^2}}e^{-\frac{x^2}{2x_0^2}}.$$

Hence the chance that this error will be less than x is

$$P = \frac{2}{\sqrt{2\pi x_0^2}}\int_0^x e^{-\frac{x^2}{2x_0^2}}dx = \frac{2}{\sqrt{\pi}}\int_0^t e^{-t^2}dt, \text{ where}$$

$$t = \frac{x}{\sqrt{2x_0^2}}.$$

Hence the Proposition is true.

Example 117. It will be seen from the tables in the back of the book that when $P = \frac{1}{2}$, $t = \cdot 476936$.

And the value of x for $P = \frac{1}{2}$ is called the Median error, and usually denoted by r.

Hence for a single measurement

$$t = \frac{r}{\sqrt{2x_0^2}} = \cdot 476936.$$

$$\text{Hence } r = \cdot 67449 \sqrt{\frac{\Sigma x^2}{n}} = \cdot 67449 x_0.$$

Since $P = \frac{1}{2}$ for the Median error it is the error that is as likely to be exceeded as not, by Definition 3. Normally there is the same number of errors smaller than r that there is larger than r. It is often called the Probable error, but as it is not at all probable this name is very misleading.

Example 118. If the errors in any 5 measurements are x_1, x_2, x_3, x_4 and x_5, what is the Median error for one measurement ?

$$\text{It is } r = \cdot 6745 x_0 = \cdot 6745 \sqrt{\frac{\Sigma x^2}{5}}.$$

What is the chance that the error some time during a series will be greater than $7r$?

It is $(\frac{1}{2})^{16} = \cdot 00002$, very nearly.

The t and P Tables in the back of the book give the value of P for all values of t. Very old tables.

$e^{-\frac{x^2}{2x_0^2}} = (\frac{1}{2})^{\frac{s^2}{3}}$, very nearly, where $s = \frac{x}{r}$, and r is the Median error at the end of the series.

As this last formula is very simple, and precise enough for almost any case, it will, no doubt, very often be used to find the chance that an error will be greater than $x = sr$ some time during a series.

And since when the error is large it is almost certain to be near the end of the series this formula may be used for the chance of an error greater than $x = sr$ at the end of the series when s is large.

$\frac{1}{5}\left(\frac{1}{2}\right)^{\frac{s^2}{3}}$ agrees very closely with the tables in the back of the book for all values of s not less than 5.

And for values of s less than 1 the formula $P = \frac{s}{2}$ is a rough approximation.

And by Propositions 11 and 13 a very handy formula that is close enough for most purposes for the chance that an error will be greater than sx_a some time during a series of games or observations is $\left(\frac{1}{2}\right)^{\frac{s^2}{2}}$. And if s is not less than 3 a very close formula for the chance that an error will be greater than sx_a at the end of a series or observation is $\frac{1}{3}\left(\frac{1}{2}\right)^{\frac{s^2}{2}}$.

If the errors are 2, 0, — 5, 0 and 6, the average error is $x_a = 2{\cdot}6$.

The average error in a line is about $\frac{4}{5}$ of the Normal error, while the Median error is about $\frac{2}{3}$ of the Normal error. And the Median error is about $\frac{5}{6}$ of the average error.

The average error $x_a = v_a\sqrt{\frac{n}{n-1}}$, where v_a is the average residual error, and n is the number of errors. See Chap. III. But in rough work the residual error may be taken for the whole error when there are many errors, as in the numerous cases where the chances of errors will be solved mentally.

In practical work the average error will be used a great deal, as it is so easy to find in linear measurements, while the median error is easiest to find in target practice.

If any set of observations is repeated the disagreement of the two Normal errors will be as great as that between the Normal and average errors. The disagreement in both cases is due only to the vagaries of chance, and if equal weight is given to each a better result than that from either will be obtained. In fact, it is the same as taking twice as many observations. We would expect the disagreement to be considerable in many cases, and the advantage of having them both should not be wasted by using only one in precise work.

For rough work these formulæ and that of Proposition 7 are all that are required And they can be reduced mentally without using any tables In this book the problems are usually worked out by the more precise formulæ. And it will be a good exercise to check them by these approximate formulæ, and them often.

Example 119. In 5 series which are the more likely errors, 3, 4, 2, 6, 1, or 2, 0, 5, 0, 6 ? And what are the Normal errors for one series or measurement ?

By Proposition 11 we must add the squares of successive errors.

The sum of the squares of the former is 66 and of the latter is 65.

Hence the latter are more likely, by Proposition 12.

The Normal errors for one series are $\sqrt{\frac{66}{5}} = 3{\cdot}633$ and $\sqrt{\frac{65}{5}} = 3{\cdot}605$, by Definition 12.

Example 120. What are the Median errors for one series ?

They are $\cdot6745\sqrt{\frac{66}{5}} = 2{\cdot}451$ and $\cdot6745\sqrt{\frac{65}{5}} = 2{\cdot}432$.

What are the average errors for one series?

They are $3\frac{1}{5}$ and $2\frac{3}{5}$.

Example 121. What are the Median errors for the whole five series?

They are $\cdot 6745\sqrt{66}$ and $\cdot 6745\sqrt{65}$.

Example 122. What is the chance of 4900 heads and 5100 tails in 10,000 flips of a coin?

$$p = q = \tfrac{1}{2}.$$

If there is no error there will be $np = 5000$ heads and $nq = 5000$ tails, by Definition 3 and Axiom 3.

Heads may be taken as positive and tails as negative.

Hence the error is $-x = -100$ heads.

Hence the chance is $\frac{1}{\sqrt{2\pi npq}} e^{-\frac{x^2}{2npq}}$

$$= \frac{1}{\sqrt{2\pi \times 10{,}000 \times \frac{1}{2} \times \frac{1}{2}}} e^{-\frac{10{,}000}{5{,}000}}$$

$$= \cdot 007979 \times \cdot 13534 = \cdot 0012.$$

Example 123. What is the chance that there will be more than 50 too many heads for that stage at some stage of the series?

It is $\frac{1}{2}e^{-\frac{x^2}{2npq}} = \frac{1}{2}e^{-\frac{2500}{5000}} = \cdot 30$. See Example 25.

Example 124. What is the chance that in 6000 throws of a die there will be 950 aces and 5050 not aces?

$$p = \tfrac{1}{6},\ q = \tfrac{5}{6}.$$

If there is no error there will be 1000 aces and 5000 not aces, by Definition 3 and Axiom 3.

The chance of an error of — 50 is

$$y = \frac{1}{\sqrt{2\pi npq}} e^{-\frac{x^2}{2npq}}$$

$$= \frac{1}{\sqrt{2 \times 6000 \times \frac{1}{6} \times \frac{5}{6}}} e^{-\frac{2500}{2 \times 6000 \times \frac{1}{6} \times \frac{5}{6}}}$$

$$= \cdot 0138 \times \cdot 2231 = \cdot 003.$$

Example 125. What is the chance that there will be more than 30 too many aces for that stage at some stage of the series ?

It is $y = \frac{1}{2}e^{-\frac{x^2}{2npq}} = \frac{1}{2}e^{-\cdot 54} = \cdot 2914.$ See Ex. 26.

Example 126. If two men flip a coin for a cent a flip what is the chance that one or the other will lose exactly 10 cents in 1000 flips ?

It is $y = \frac{2}{\sqrt{2\pi npq}} e^{-\frac{\left(\frac{x}{a+b}\right)^2}{2npq}}$

$$= \cdot 0505 \times \cdot 951 = \cdot 048.$$

We must divide 10 by the sum of the stakes to get the yes and no. Check result by Ex. 25.

Example 127. What is the chance that one or the other will lose more than 10 some time during 1000 flips ?

It is $y = e^{-\frac{\left(\frac{x}{a+b}\right)^2}{2npq}} = e^{-\frac{25}{500}} = \cdot 95.$

Yes or no is only pronounced every time the sum of the stakes is won or not won. Check result by Ex. 25.

Hence to get the number of pairs of contradictories we must take $\frac{x}{a+b}$ and not x.

For convenience we may call x the cash error, but the real error is $\frac{x}{a+b}$, by Axiom 3. However, in dealing with money it is simpler to have a cash error, though not exact. But the result is the same.

Example 128. If in any country the chance of blindness is $p_1 = \cdot 001$ and of deafness is $p_2 = \cdot 01$ and that of dumbness is $p_3 = \cdot 1$, what is the chance that in a district of 1,000,000 population there will be exactly 100,000 who are neither blind nor deaf but are dumb ?

$n = 1{,}000{,}000$, $p = q_1\, q_2\, p_3 = \cdot 098901$, $q = 1 - \cdot 098901 = \cdot 901099$, $np = 98901$, $np + x = 100{,}000$, by Definitions 3 and 11.

Hence $x = 100{,}000 - 98901 = +\,1099$, by Definition 11.

$$y = \frac{1}{\sqrt{2\pi npq}}\, e^{-\frac{x^2}{2npq}} = \cdot 00134 \times \cdot 00115 = \cdot 000{,}002,$$

which is the answer.

Example 129. What is the chance that the error will be less than 1099 ?

$$t = \frac{x}{\sqrt{2npq}} = 2\cdot 60.$$

Hence $P = \cdot 9998$.

That is, the chance that there will be between 98901 and 101099 is ·9998, which is the answer.

Example 130. What is the chance that there will be between 98901 and 100,000?

It is $\frac{1}{2}$ of ·9998 = ·4999.

Example 131. What is the chance that the error will be more than 1099?

It is 1 — ·9998 = ·0002.

Example 132. What is the chance that there will be less than 98901?

It is $\frac{1}{2}$ of ·0002 = ·0001.

Example 133. What is the chance that there will be more than 101099?

It is also $\frac{1}{2}$ of ·0002 = ·0001.

Example 134. Two men each have \$10 and flip a coin for \$1 a flip. What is the chance that one or the other will be ruined some time during 100 flips?

$$a + b = 2.$$

It is the chance of a cash error of more than 10 some time during 100 trials.

$$\text{Hence } y = e^{-\frac{\left(\frac{x}{a+b}\right)^2}{2npq}} = e^{-\frac{25}{2 \times 100 \times \frac{1}{2} \times \frac{1}{2}}} = \cdot 6065.$$

Example 135. This finishes Chapter I, which is composed of nothing but definitions, axioms and propositions with a few examples to elucidate them. If the student learns this chapter well he will have little difficulty with anything that follows, for it is all based on this.

In this book the law of change is worked out from *a priori* axioms, and not from axioms taken from experience, which is a departure from the usual method.

And these same axioms are evidently those of evolution, which undoubtedly is the result of the interference of yes and no, as error is. Evolution is error. If yes and no did not interfere in nature we would always have the expectation, by Definition 11 and Axiom 3, and there would be no change in species. And we would expect that the law governing the actions of men would govern the rest of nature.

Thus by measuring the change in a thousand years it would be possible to get an infinitely closer value for the age of the world than in any other way. Or the time it took one species to change from another.

This subject is of infinite utility because it deals with the combinations that are taking place around us.

Such cases are infinitely complicated, but this subject deals with infinitely complicated cases better than any other. And with things about which we know very little.

Out of any million men we would not expect all to be equally lucky. We would expect one to be very unlucky and one to be very lucky. And a gradual difference from one to the other.

The unlucky man would think that it was something in himself that made him so unlucky and the other would think that he made his good luck. But it is simply because they belong to a group of a million men all of whom naturally have different degrees of luck. And this takes place when all the men are equally circumspect, diligent and careful. If they are not so there is a constant error or expectation that has nothing to do with luck. Luck is only the accidental changes from the expectation, and these errors follow the law of combinations. Certain combinations occur once in a million. This combination strikes the one man in a million.

Another combination once in a thousand. This one strikes a thousand men out of the million. Another occurs once in ten times. This one strikes a hundred thousand men out of the million.

But the expectation is proportional to the number of trials, while the error is proportional to the square root of the number of trials.

Hence if a man leads a good life his expectation is so great in comparison with his error that it is almost negligible. And he is largely independent of luck.

The rest of the book applies the laws of chance and error to the daily affairs of life, and shows how to make use of every scrap of knowledge that we may be able to gather either from our own labour or from that of others, and to make a little knowledge go a long way.

Chapter II

GAMES WHOSE EXPECTATION IS 0

Example 136. We have shown that the number of combinations from n things when all the elements are very small is e^n.

Suppose that there are r combinations, how many things does it require to produce them?

The number of things must be n where $r = e^n$. The n things are called the Natural log of r, and written $\log_e r = n$, and it is said that n is the log of r to the base e.

In the common system of notation there are 10 numerals, 1, 2, 3, 4, 5, 6, 7, 8, 9 and 0. Hence any number must be some power of 10. Thus $100 = (10)^2$, $1000 = (10)^3$, etc

Hence in dealing with numbers it is simpler to take 10 for the base than e. The base of Common logs is 10.

Thus r combinations are produced by $\log_e r$ things. Hence $e^{\log_e r} = r$, by Proposition 2, where r can have any value.

Hence $e^{\log_e 10} = 10$.

Hence $e^{n \log_e 10} = 10^n = a$.

And since $10^n = a$, and 10 is the base of Common logs we have $n = \log_{10} a$.

And since $e^{n \log_e 10} = a$, we have

$\log_e a = n\log_e 10$, by Proposition 2,

$= \log_{10} a \times \log_e 10$, where a can have any value.

Put $a = e$.

Hence $\log_e e = \log_{10} e \log_e 10$.

But $\log_e e$ is the number of things required to produce e combinations.

Hence $\log_e e = 1$, by Proposition 2.

Hence $1 = \log_{10} e \log_e 10$.

Any table of Common logs gives $\log_{10} e = \cdot 4343$.

Hence $\log_e 10 = \frac{1}{\cdot 4343} = 2\cdot 3026$.

Hence $\log_e a = 2\cdot 3026 \log_{10} a$, and $\log_{10} a = \cdot 4343 \log_e a$.

Since chance is purely a matter of things and combinations by Axiom 10 and Definition 1, 2, 3 and 4, and as Natural logs are the things, we must expect to have to use Tables of logs a great deal in chance. The tables are very easy to use.

It is very easy to multiply numbers by adding their logs since the sum of the logs is the log of the product.

If $a = (10)^{y_1}$, and $b = (10)^{y_2}$,

$a \times b = (10)^{y_1 + y_2} = (10)^{\log_{10} a + \log_{10} b}$.

Hence $\log_{10} ab = \log_{10} a + \log_{10} b$.

If $a = 5$ and $b = 9$, we have

$\log_{10} ab = \log_{10} 5 + \log_{10} 9$.

The tables of Common logs give

$\log_{10} 5 = \cdot 69897$

$\log_{10} 9 = \cdot 95424$.

Hence $\log_{10} ab = \cdot 69897 + \cdot 95424 = 1\cdot 65321$.

The tables of logs show that $1\cdot 65321$ is the log of 45.

When we are using Common logs we will seldom put the base, but simply log 45 instead of $\log_{10} 45$. But when we use Natural logs we will always put the base, thus, $\log_e 45$.

Since Common logs are easier to use with our system of notation than Natural logs they will nearly always be used.

Example 137. Suppose that A has only 25 cents and he flips a coin for a cent a flip. If head turns up he gets a cent, and if tail he gives a cent. What is the chance that he will lose the 25 cents some time during 400 flips ?

His chance to win is $p = \frac{1}{2}$ and his chance to lose is $q = \frac{1}{2}$ in any trial. The sum of the stakes is $a + b = 2$.

Hence $\frac{x}{a + b} = 12{\cdot}5$.

The chance that he will lose the 25 cents some time during 400 trials is

$$y = \tfrac{1}{2}e^{-\frac{\left(\frac{x}{a+b}\right)^2}{2npq}} = \tfrac{1}{2}e^{-{\cdot}78}$$, by Proposition 12 and Definition 14.

$$\log y = -\log\ 2 - {\cdot}78 \times {\cdot}4343$$
$$= -{\cdot}63978 = \bar{1}{\cdot}36022.$$

Hence $y = {\cdot}23$, or about $\frac{1}{4}$.

The 25 cents is called his fortune, and when it is lost he is said to be ruined.

Hence the odds are about 1 to 3 that he will be ruined some time during the series of 400 trials.

Example 138. If his adversary B also has 25 cents what is the chance that one or the other will be ruined some time during the 400 flips ?

It is the chance of an error of x or more some time during the series of 400 trials, or $y = e^{-\frac{\left(\frac{x}{a+b}\right)^2}{2npq}} = {\cdot}46$, or about $\frac{1}{2}$, by Proposition 12 and Definition 14.

The odds are about 1 to 1 that one or the other will be ruined some time during the series of 400 trials.

Example 139 If his adversary B has a dollar, what is the chance that B will be ruined sometimes during the series of 400 trials?

If A's limiting error is x B's is $rx = 4x$.

Hence $y = \frac{1}{2}e^{-\frac{\left(\frac{rx}{a+b}\right)^2}{2npq}} = \frac{1}{2}\left(e^{-\frac{\left(\frac{x}{a+b}\right)^2}{2npq}}\right)^{r^2}$

$= \frac{1}{2}(\cdot 46)^{r^2} = \frac{1}{2}(\cdot 46)^{16} = \cdot 000{,}02$, by Definition 14.

B's game is very safe, but A's is not.

Example 140. Or take 4 cents each and a series of 10 trials.

If they each have a fortune of 25 cents the chance that one or the other will be ruined is $y = e^{-\frac{\left(\frac{x}{a+b}\right)^2}{2npq}}$.

If $\frac{\left(\frac{x}{a+b}\right)^2}{n}$ is constant y will always have the same value.

$\left(\frac{x}{a+b}\right)^2 = (12\cdot 5)^2 = 156\cdot 25.$

$n = 400.$

If we divide each of these by 40 we get $n = 10$ and $\left(\frac{x}{a+b}\right)^2 =$ about 4.

Hence $\frac{x}{a+b} = 2$, and $x = 4$, which is the fortune.

And the chance that one or the other will be ruined some time during the series of 10 trials is the same as when the fortunes were each 25 cents and n was 400, except that as the elements of the series are not so small the result will not be quite so precise, by Proposition 2.

The chance that A will be ruined is $\frac{1}{2}$ of $\frac{1}{2} = \frac{1}{4}$.

The chance that one or the other will be ruined is twice the chance that he will be ruined.

To win and to fail have opposite signs. Hence when we say that an error is less than x we mean that it is between 0 and x on either side of 0. An error of $-x$ is an error of x measured in the minus direction. An error of $+x$ is an error of x measured in the plus direction. An error of -13 is greater than an error of $+12{\cdot}5$ because it is farther from 0.

When we speak of an error of x we mean that it may be measured in either direction from 0.

When we speak of an error greater than $-x$ we mean an error measured in the minus direction and numerically greater than x, and an error greater than $+x$ is an error measured in the plus direction and numerically greater than x. An error greater than x means an error measured in either direction from 0 and numerically greater than x.

Example 141. If A's fortune is 4 cents, what is the chance thay he will have exactly the same at the end of a series of 10 trials as he had at the beginning if he can borrow enough money to tide over the unlucky parts till the end of the series ?

It is the chance of no error at the end of the series or $\frac{1}{\sqrt{2\pi npq}} = {\cdot}25$, or about $\frac{1}{4}$. It is also the chance of no gain or the chance of no loss.

The odds are about 1 to 3 that he will neither gain nor lose anything.

Example 142. What is the chance that he will gain exactly 1 cent at the end of the series?

$$x = 1, \frac{x}{a+b} = \cdot 5$$

$$y = \frac{1}{\sqrt{2\pi npq}} e^{-\frac{\left(\frac{x}{a+b}\right)^2}{2npq}} = \cdot 21 \text{ or about once in 5 series}$$

of 10 trials each, by Proposition 12 and Definition 14.

Example 143. If we want to know the chance that the error will be less than 5 cents at the end of a series of 400 trials in flipping a coin for a cent a flip we have

$$t = \frac{\frac{x}{a+b}}{\sqrt{2npq}} = \frac{2\cdot 5}{\sqrt{2 \times 400 \times \frac{1}{2} \times \frac{1}{2}}} = \cdot 1768, \text{ by}$$

Proposition 13.

For $t = \cdot 1768$ the Tables in the back of the book give $P = \cdot 1974$.

That is, the chance that the error will be less than 5 is ·1974, which is the chance that A will either gain or lose less than 5 cents at the end of 400 flips if he is playing a cent a flip.

Example 144. The chance that the error will be greater than 5 is $1 - \cdot 1974 = \cdot 8026$, by Definition 3.

Example 145. The chance that the error will be less than $+ 5$, or that the gain will be less than 5 cents is $\frac{1}{2}$ of $\cdot 1974 = \cdot 0987$.

The chance that the loss will be less than 5 cents is the same.

Example 146. The chance that the gain will be more than 5 cents is $\frac{1}{2}$ of ·8026 = ·4013.

The chance that the loss will be more than 5 cents is the same.

This is at the end of the series and not some time during the series. It is supposed that he has enough money to tide over the bad luck in parts of the series.

Example 147. What is the chance that in 10 trials at Head or Tail there will be 4 heads and 6 tails ?

$$p = q = \tfrac{1}{2}.$$

If there is no error there will be $np = 5$ heads and $nq = 5$ tails.

We may give the direction + to heads and — to tails.

Hence the error is $4 - 5 = -1 = -x$.

The chance that the error will be — 1 at the end of the series is $y = \frac{1}{\sqrt{2\pi npq}} e^{-\frac{x^2}{2npq}}$

$$= \frac{1}{\sqrt{2\pi \times 10 \times \frac{1}{2} \times \frac{1}{2}}} e^{-\frac{1^2}{2 \times 10 \times \frac{1}{2} \times \frac{1}{2}}}$$

$\log y = -\frac{1}{2} \log (2\pi \times 10 \times \frac{1}{2} \times \frac{1}{2}) - \frac{1}{2 \times 10 \times \frac{1}{2} \times \frac{1}{2}}$ $\log e$.

Hence y = ·2523 × ·8187 = ·2, or about once in 5 series. Check result by Ex. 25.

The chance that there will be either 4 heads or 4 tails is ·2 × 2 = ·4, or about twice in 5 series.

Example 148. What is the chance that there will be an error of at least 3 some time during the series ?

It is $e^{-\frac{x^2}{2npq}} = \cdot 16$, or about once in 6 series.

Example 149. What is the chance of an error of — 3 or more at some stage of the series?

It is $\frac{1}{2}e^{-\frac{x^2}{2npq}} = \frac{1}{2}$ of $\cdot 16 = \cdot 08$, or about once in 12 series.

Example 150. What is the chance that the error will be less than 5 at the end of a series of 100 trials ?

$$t = \frac{x}{\sqrt{2npq}} = \frac{7}{10} = \cdot 7.$$

For $t = \cdot 7$ the Tables in the back of the book give $P = \cdot 68$, which is the answer.

Example 151. What is the chance that the error will be less than — 5 at the end of the series, that is, between 0 and — 5 ?

It is $\frac{1}{2}$ of $\cdot 68 = \cdot 34$.

Example 152. What is the chance that the error will be greater than 5 ?

It is $1 - \cdot 68 = \cdot 32$.

The error must be less than x wins.
or greater than x ,,
or less than x failures.
or greater than x ,,

And wins may be taken as positive and failures as negative.

Hence the sum of these four chances is 1, by Definition 3.

And $+ x$ and $- x$ are equally likely when the elements are all very small, by Proposition 11.

Hence the chance that the error will be + is ½.
And the chance that the error will be — is ½.
The chance that it will be less than 5 is ·68.
The chance that it will be less than + 5 is ½ of ·68= ·34.
The chance that it will be less than — 5 is the same.
The chance that it will not be less than + 5 is 1 — ·34 = ·66.
The chance that it will not be less than — 5 is the same.
The chance that it will be greater than 5 is 1 — ·68 = ·32.
The chance that it will be greater than + 5 is ½ of ·32 = ·16.
The chance that it will be greater than — 5 is the same.
The chance that it will not be greater than + 5 is 1 — ·16 = ·84.
The chance that it will not be greater than — 5 is the same.

It is assumed in the above that the chance that the error will be exactly x is so small that it may be neglected. When it is not we can apply a correction, which is seldom necessary, by Definition 14.

Example 153. What is the chance that the error will be less than + 5 ?

It is ½ of ·68 = ·34. The chance that the error will be less than + 5 is the same as the chance that it will be less than — 5.

Example 154. What is the chance that the error will be minus ?

It is ½ since it is equally likely to be plus or minus, and the sum of the chances must be 1, by Definition 3.

Example 155. If A tries to throw ace at the game of

Dice for 1200 trials, what is the chance of an error of more than 20 some time during the series ?

A die has 6 sides, one of which is ace and 5 are not ace.

Hence $p = \frac{1}{6}$ and $q = \frac{5}{6}$.

The chance is $y = e^{-\frac{x^2}{2npq}} = e^{-\frac{400}{333}} = \cdot 3.$

Example 156. What is the chance of an error of more than — 20 some time during the series ?

It is $\frac{1}{2}e^{-\frac{x^2}{2npq}} = \cdot 15$, by Proposition 13.

Example 157. What is the chance of more than 220 aces at the end of the series of 1200 trials ?

If there is no error there will be $np = 200$ aces and $nq = 1000$ not aces.

Hence the error is $220 - 200 = +20$.

$$t = \frac{x}{\sqrt{2npq}} = \frac{20}{\sqrt{2 \times 1200 \times \frac{1}{6} \times \frac{5}{6}}} = 1 \cdot 09, \text{ by Pro. 13.}$$

Hence P = ·877.

This is the chance that the error will be less than 20.

Hence the chance that the error will be more than 20 is 1 — ·877 = ·123.

Hence the chance that the error will be more than + 20 is $\frac{1}{2}$ of ·123 = ·06, which is the chance of more than 220 aces at the end of the series of 1200 trials.

When n is equal to 32 the following Table gives the Natural and the Common chances of an error of x for the game of Head or Tail. The formulæ used are

$$\frac{2}{\sqrt{2\pi npq}} e^{-\frac{x^2}{2npq}} \text{ and } \frac{2\,\underline{|n}}{\underline{|np+x}\;\underline{|nq-x}} p^{np+x} q^{nq-x}$$

And $p = q = \frac{1}{2}$. See Definition 19.

x	NATURAL CHANCE.	COMMON CHANCE.
0	·14105	·13995
1	·26500	·26344
2	·21970	·21952
3	·16073	·16176
4	·10378	·10514
5	·05913	·06008
6	·02977	·03004
7	·01319	·01306
8	·00517	·00490
9	·00179	·00157
10	·00055	·00042
11	·00015	·00009
12	·000035	·000017
13	·000007	·0000023
14	·0000014	·00000023
15	·00000022	·000000015
16	·00000003	·0000000005

Thus when npq is so small as 8 there is not much difference between Natural and Common chance for small errors. As $\sqrt{npq}$ becomes larger the difference practically disappears, because the elements of the normal error become very small.

Since a small error occurs very much oftener than a large one, and the agreement is close for small errors, the chances for an error less than x agree very much better. The large errors have very little weight then. An error of 1 occurs about a billion times as often as an error of 16 in the above Table.

An error of x or greater is not the same as an error greater than x, because x is included in the former and excluded in the latter. But if $\sqrt{npq}$ is very large they may be taken as the same, since when npq is very large the chance of an error of exactly x is likely very small. Hence when npq is very large we may consider that the chances of an error less than x and greater than x each include one half of the chance of an error of exactly x. And if the chance of an error less than x is P the chance of an error greater than x is $1 - P$, by Definitions 3 and 14.

But since the chance of an error of x at the end of a series of n trials is $\frac{2}{\sqrt{2npq}}e^{-\frac{x^2}{2npq}}$ the chance of an error less than x is $P - \frac{1}{\sqrt{2\pi npq}}e^{-\frac{x^2}{2npq}}$, and the chance of an error greater than x is $1 - P - \frac{1}{\sqrt{2\pi npq}}e^{-\frac{x^2}{2npq}}$.

This makes a considerable difference when $\sqrt{npq}$ is small. But for simplicity in the examples we have assumed that the chance of an error less than x and that of an error greater than x each include half the chance of an error of x.

Example 158. Take the example what is the chance of an error greater than 5 in 100 trials at Head or Tail?

The answer we got for this in Example 152 was ·32.

But $\frac{1}{\sqrt{2\pi npq}}e^{-\frac{x^2}{2npq}} = \cdot 05.$

Hence the chance of an error greater than 5 is only $\cdot 32 - \cdot 05 = \cdot 27$.

Hence unless npq is very large this correction should be applied if great precision is required.

Chapter III

DIRECT OBSERVATIONS

Example 159. Since the most likely error at the end of a series is 0, by Proposition 12, if anything is measured a series of n times under the same conditions, and the mangitudes got are l_1, l_2, l_3, and so on, respectively, the most likely magnitude is $\frac{l_1 + l_2 + l_3 + \text{etc.}}{n} = l_0$, or the average magnitude.

And the *residual* errors are

$v_1 = l_1 - l_0$, $v_2 = l_2 - l_0$, $v_3 = l_3 - l_0$, and so on.

And $l_1 + l_2 + l_3 + \text{etc.} = nl_0$.

If the errors are x_1, x_2, x_3, and so on, the total number of chances of a pair of contradictories is $2\pi x_1^2$, $2\pi x_2^2$, $2\pi x_3^2$ and so on, respectively, by Proposition 11.

And there cannot be any contradictories between different measurements, by Definition 8.

Any measurement is called an observation, and as *obs* is very simple to write it will generally be used for measurement, measurements, measure, measured, observe, observed, observation, etc.

Hence the total number of Chances of a pair of contradictories in the series of n *obs* is $2\pi\ (x_1^2 + x_2^2 + x_3^2 + \text{etc}) = 2\pi\Sigma x^2$.

Hence the number of chances of a pair of contradictories for the average is $\frac{2\pi\Sigma x^2}{n^2} = \frac{2\pi x_0^2}{n}$.

Hence the weight of the average *obs* is $p = n$, by Definition 13.

The weight is usually denoted by p.

Now suppose that these n *obs* of the same thing are not made under the same set of conditions, but that the weights of the *obs* are p_1, p_2, p_3, etc , and the normal errors are x_1, x_2, x_3, etc., respectively.

Hence $p_1 x_1^2 = p_2 x_2^2 = p_3 x_3^2 =$ etc., by Definition 13.

Hence the weights are inversely proportional to the squares of the Normal errors.

And the weight of the average, l_0, is $p_1 + p_2 + p_3 +$ etc. $=$ P.

Hence if x_0 is the Normal error for an *obs* of weight 1, the Normal error of the average, l_0, is $\frac{x_0}{\sqrt{P}}$.

And $Px_0^2 = p_1 x_1^2 = p_2 x_2^2 = p_3 x_3^2 =$ etc. $= k$.

And since $P = p_1 + p_2 + p_3 +$ etc.

$$\frac{k}{x_0^2} = \frac{k}{x_1^2} + \frac{k}{x_2^2} + \frac{k}{x_3^2} + \text{etc.}$$

And $\frac{1}{x_0^2} = \frac{1}{x_1^2} + \frac{1}{x_2^2} + \frac{1}{x_3^2} +$ etc.

And since the Median error is proportional to the normal error the Median error of the average magnitude is r_0, where

$$\frac{1}{r_0^2} = \frac{1}{r_1^2} + \frac{1}{r_2^2} + \frac{1}{r_3^2} + \text{etc.}$$

where r_1, r_2, r_3, etc., are the Median errors of the respective series.

Since the average magnitude itself has an error the residual errors do not include all the error, but only what is left of it.

In a series of *obs* of anything all made under the same set of conditions, the number of combinations of contradictories is $2\pi\,\Sigma\,x^2$, and for a single *obs* is $2\pi\,\frac{\Sigma x^2}{n} = 2\pi x^2_0$, where x_0 is the Normal error for a single *obs*.

And since the weight of the average is n times the weight of a single *obs* the number of chances of a pair of contradictories of the average magnitude is

$$\frac{2\pi x_0{}^2}{n} = 2\pi\,\frac{x_0{}^2}{n}\,.$$

And the Normal error of the average magnitude is

$$\sqrt{\frac{x_0{}^2}{n}} = \frac{x_0}{\sqrt{n}}\,.$$

And each residual has its contradictory, and the error of the average has its contradictory, by Axiom 3.

And there are no other contradictories, by Definition 8 and 14.

The number of chances of a pair of the residual contradictories for one *obs* is

$$2\pi\,\frac{v_1{}^2 + v_2{}^2 + v_3{}^2 + \text{etc.}}{n} = 2\pi\,\frac{\Sigma v^2}{n} = 2\pi v_0{}^2.$$

And the number of chances of a pair of contradictories for the average is $2\pi\frac{x_0{}^2}{n}$.

And the total number of chances of a pair of contradictories for one *obs* is $2\pi\,\frac{\Sigma x^2}{n} = 2\pi x_0{}^2$

Hence $2\pi x_0{}^2 = 2\pi\,\frac{\Sigma\,v^2}{n} + 2\pi\,\frac{x_0{}^2}{n}\,.$

Hence $x_0^2 = \dfrac{\Sigma v^2}{n-1}$.

Hence $x_0 = \sqrt{\dfrac{\Sigma v^2}{n-1}}$.

And the Normal error of the average magnitude is

$$\sqrt{\frac{\Sigma v^2}{n\,(n-1)}}$$

And $\Sigma x^2 = n x_0^2 = \dfrac{n}{n-1}\,\Sigma v^2$.

Suppose that a line is measured 4 times with the same tape and care and the lengths got are

1000·23
999·43
999·27
1000·43

The average length is $\dfrac{3999{\cdot}36}{4} = 999{\cdot}84$.

The residuals are

+ ·39
— ·41
— ·57
+ ·59

It will be noticed that the sum of the residuals is 0. The squares of the residuals are

·1521
·1681
·3249
·3481

$$\frac{\Sigma v^2}{n-1} = \frac{{\cdot}9932}{3} = {\cdot}3311.$$

$$x_0 = \sqrt{\frac{\Sigma v^2}{n-1}} = \cdot 5755.$$

The chance of a total error of x or greater some time during n *obs* is

$$y = e^{-\frac{x^2}{2n\frac{\Sigma v^2}{n-1}}} = e^{-\frac{x^2}{\cdot 6622 \times 4}}$$

Example 160. What is the chance of an error of less than ·5 at the end of a single *obs* ?

$$t = \frac{x}{\sqrt{2\frac{\Sigma v^2}{n-1}}} = \cdot 62, \text{ by Proposition 13.}$$

For $t = \cdot 62$ the Tables give P = ·62, which is the chance of an error of less than ·5.

The chance of an error of more than ·5 is 1 — ·62 = ·38. And since an error of measurement changes by very small elements there is no correction required, since the chance of an error of exactly x is 0.

Example 161. What is the chance that the error will be x or greater some time during r *obs* ?

It is $y = e^{-\frac{x^2}{2r\frac{\Sigma v^2}{n-1}}} = e^{-\frac{x^2}{\cdot 6622 r}}$, by Proposition 13.

Example 162. What is the chance that there will be a total error of 1·5 or greater some time during 2 *obs* ?

It is $y = e^{-\frac{(1\cdot 5)^2}{\cdot 6622 \times 2}} = \cdot 1829.$

Example 163. What is the chance that there will be a total error of ·5 or greater at the end of 3 *obs* ?

$$t = \frac{x}{\sqrt{2r\frac{\Sigma v^2}{n-1}}} = \frac{\cdot 5}{\sqrt{6 \times \cdot 3311}} = \cdot 35.$$

Hence P = ·38.

Hence 1 — P = ·62, which is the answer.

Example 164. What is the chance that the error will be ·8 or greater some time during one *obs* ?

It is $y = e^{-\frac{x^2}{2r\frac{\Sigma v^2}{n-1}}} = e^{-\frac{\cdot 64}{\cdot 6622}} = \cdot 38.$

The Normal error of the average length is

$$\frac{\cdot 5755}{\sqrt{n}} = \cdot 2878.$$

The Median error of the average length is $\frac{\cdot 388}{\sqrt{n}} = \cdot 194$

The average length of the above line was 999·84.

It is usual to write the Median error of the average length after the average length, thus, 999·84 ± ·194 is the Median length of the line.

That is the true length of the line is as likely to be between 1000·034 and 999·646 as not.

Example 165. If A has measured an angle 4 times with the same instrument and care and got

20° 14′·3
20° 14′·7
20° 13′·9
20° 13′·5

what is the Median error ?

The average angle is 20° 14′·1.

$$v_1 = +\cdot 2$$
$$v_2 = +\cdot 6$$
$$v_3 = -\cdot 2$$
$$v_4 = -\cdot 6$$

$\Sigma v = 0$, as it should.

$$v_1^2 = \cdot 04$$
$$v_2^2 = \cdot 36$$
$$v_3^2 = \cdot 04$$
$$v_4^2 = \cdot 36$$

$\Sigma\, v^2 = \cdot 80.$

$$\frac{\Sigma\, v^2}{n-1} = \cdot 2667.$$

The Normal error for one *obs* is

$$\sqrt{\frac{\Sigma x^2}{n}} = \sqrt{\frac{\Sigma v^2}{n-1}} = \cdot 5164.$$

The Normal error of the average angle is $\dfrac{\cdot 5164}{\sqrt{4}} = \cdot 2582$

The Median error for one *obs* is ·6745 of ·5164 = ·3483.

The Median error of the average angle is ·1742.

The Median angle is 20° 14′·1 ± 0′·1742.

That is it is as likely as not to be between 20° 14′·2742 and 20° 13′·9258.

Example 166. What is the chance that an error will be between ·5 and ·6 in the above line?

We may find the chance that an error will be less than ·5 and then the chance that it will be less than ·6 and subtract the former chance from the latter, which will give the chance that an error will be between ·5 and ·6.

For the first $t = \dfrac{x}{\sqrt{2\dfrac{\Sigma\, v^2}{n-1}}} = \cdot 6144.$

The Tables give for $t = \cdot 6144$, P $= \cdot 6151$.

Similarly the chance that an error will be less than ·6 is ·7029.

Hence the chance that an error will be between ·5 and ·6 is ·7029 — ·6151 = ·0878.

Since ·6 — ·5 is a small element of the average ·55 we can work this by the formula for the Natural chance of an error of ·55, by Definition 19.

The chance of an element c of error at $x = \cdot 55$ is

$$y = \frac{2c}{\sqrt{2\pi x_0^2}} e^{-\frac{x^2}{2x_0^2}} = \frac{2\,(\cdot 6 - \cdot 5)}{\sqrt{2\pi \times \cdot 3311}} e^{-\frac{x^2}{\cdot 6622}} = \cdot 0878$$

the same as before.

Example 167. If A has taken the elevation of a Bench Mark 4 times with the same instrument and care and got the elevations

100·34
99·67
100·21
99·86

what is the Median elevation ?

The average elevation is 100·02.

$$v_1 = + \cdot 32$$
$$v_2 = - \cdot 35$$
$$v_3 = + \cdot 19$$
$$v_4 = - \cdot 16$$

The Normal error for a single *obs* is

$$\sqrt{\frac{\Sigma v^2}{n-1}} = \cdot 3090.$$

The Normal error for the average elevation is

$$\frac{\cdot 3090}{\sqrt{4}} = \cdot 1545.$$

The Median error for one *obs* is

$$\cdot 6745 \text{ of } \cdot 3090 = \cdot 2084.$$

The Median error of the average elevation is

$$\frac{\cdot 2084}{\sqrt{4}} = \cdot 1042.$$

The Median elevation is $100{\cdot}02 \pm \cdot 1042$.

That is it is as likely as not to be between 99·9158 and 100·1242.

Example 168. If the numerical sum of all the errors is Σx, since the same law governs the residuals that governs the other errors, we have, by Proposition 11,

$$\left(\frac{\pi\Sigma v}{n}\right)^2 = 2\pi\frac{\Sigma v^2}{n}$$

$$\frac{(\pi\Sigma v)^2}{n} = 2\pi\Sigma v^2$$

$$\frac{(\pi\Sigma v)^2}{n(n-1)} = 2\pi\frac{\Sigma v^2}{n-1}$$

$$\frac{\pi(\Sigma v)^2}{n(n-1)} = 2\frac{\Sigma v^2}{n-1}$$

$$\pi = 2n\frac{\Sigma x^2}{(\Sigma x)^2} = 2n\frac{\Sigma v^2}{(\Sigma v)^2}.$$

And it is immaterial which expression is used in any of the formulæ.

$$\text{Hence } x_0 = \sqrt{\frac{\Sigma v^2}{n-1}} = \sqrt{\frac{\pi}{2}}\cdot\frac{\Sigma v}{\sqrt{n(n-1)}} = 1{\cdot}2533\frac{\Sigma v}{\sqrt{n(n-1)}}.$$

The Median error is

$$r = \cdot 6745\sqrt{\frac{\Sigma x^2}{n}} = \cdot 8453\frac{\Sigma x}{n} = \cdot 8453 x_a,$$

$$= \cdot 6745\sqrt{\frac{\Sigma v^2}{n-1}} = \cdot 8453\frac{\Sigma v}{\sqrt{n(n-1)}}.$$

When n is very large we will use the formulæ containing Σx or Σv instead of those containing Σx^2 or Σv^2 since they are much more convenient, and when n is very large the results are practically the same.

The two methods are equally precise, and are good checks on each other.

Example 169. If a line is measured 4 times under the same set of conditions and the lengths got are

$$\begin{array}{c} 1000{\cdot}4 \\ 1000{\cdot}2 \\ 1000{\cdot}6 \\ 1000{\cdot}8 \end{array}$$

the average length is 1000·5.

The residuals are

$$\begin{aligned} v_1 &= -\cdot 1 \\ v_2 &= -\cdot 3 \\ v_3 &= +\cdot 1 \\ v_4 &= +\cdot 3 \\ \Sigma v^2 &= \cdot 2 \end{aligned}$$

$$x_0 = \sqrt{\frac{\Sigma v^2}{n-1}} = \cdot 258.$$

The Normal error of the average length is

$$\frac{\cdot 258}{\sqrt{4}} = \cdot 129.$$

The Median error of the average length is

$$\cdot 6745 \times \cdot 129 = \cdot 087.$$

The Median length of the line is 1000·5 ± ·087.

Example 170. Now suppose that it is measured 5 times under a different set of conditions and the lengths got are

1000·15
1000·26
1000·37
1000·02
1000·20

The average length is 1000·20.

The residuals are

$$v_1 = -\cdot05$$
$$v_2 = +\cdot06$$
$$v_3 = +\cdot17$$
$$v_4 = -\cdot18$$
$$v_5 = \cdot00$$
$$\Sigma v^2 = \cdot0674$$

$$x_0 = \sqrt{\frac{\Sigma v^2}{n-1}} = \cdot130.$$

The Normal error of the average length is

$$\frac{\cdot130}{\sqrt{5}} = \cdot058.$$

The Median error of the average length is

·6745 of ·058 = ·039.

The Median length of the line is

1000·20 ± .039.

Example 171. Now if we want to combine these two results to get a better result than either we have

$$\frac{1}{R^2} = \frac{1}{r_1^2} + \frac{1}{r_2^2} = \frac{1}{(\cdot087)^2} + \left(\frac{1}{\cdot039}\right)^2.$$

Hence R = ·036.

The weighted average length of the line is

$$\frac{p_1 l_1 + p_2 l_2}{p_1 + p_2} = \frac{l_1 + \frac{p_2}{p_1} l_2}{1 + \frac{p_2}{p_1}} = \frac{l_1 + \frac{r_1^2}{r_2^2} l_2}{1 + \frac{r_1^2}{r_2^2}}$$

$$= \frac{1000{\cdot}5 + 1000{\cdot}21\left(\frac{{\cdot}087}{{\cdot}039}\right)^2}{1 + \left(\frac{{\cdot}087}{{\cdot}039}\right)^2} = 1000{\cdot}25.$$

Hence the Median length of the line is 1000·25 $\pm$ ·036. And this is better than either of the others since we use more information in getting it, and the chance of anything changes as we know more about it, by Definition 1.

Example 172. If the total error for n angles measured under the same set of conditions is E the most likely error for each angle is $\frac{E}{n}$, since these errors must be all the same, and their sum must equal the total error.

If the closing error of the angles of a triangle is + 3′, the most likely error for each angle is $\frac{+3'}{3} = 1'$, for the same reason.

Example 173. If the error in a line 1000 feet long is — ·6 the most likely error for each 100 of it is $-\frac{{\cdot}6}{10} =$ — ·06, since they must all be alike, and the sum of the errors must equal the total error. The most likely error for each 100 is not the same as for any 100.

Example 174. If a line is composed of two parts which when measured give 1000·0 and 2000·3, and then the whole line is measured and gives 3000·6, the discrepancy is ·3, which is a true error, and not a residual error, since it should be 0, and must be apportioned among the 3 lines proportionally to their length, since the most likely error for each unit length must be the same, and their sum must equal the total error, or

$\frac{1}{6}$ of $\cdot 3 = \cdot 05$.
$\frac{2}{6}$ of $\cdot 3 = \cdot 10$.
$\frac{3}{6}$ of $\cdot 3 = \cdot 15$.

Hence the most likely lengths are

$1000{\cdot}0 + \cdot 05 = 1000{\cdot}05$.
$2000{\cdot}3 + \cdot 10 = 2000{\cdot}40$.
$3000{\cdot}6 - \cdot 15 = 3000{\cdot}45$.

Example 175. If the two parts of an angle are separately measured and then the whole angle measured and the discrepancy is E, it must be apportioned equally among the 3 angles. If the 2 parts are 20° 14′ and 35° 11′ and the whole angle is 55° 28′ the discrepancy is 3′, which must be approtioned 1′ to each angle, so that the most likely values of the angles are 20° 15′, 35° 12′. and 55° 27′, if they were all measured under the same set of conditions, by Definition 13.

When we measure the three angles of a triangle we know the closing true error since the sum of the three angles must be 180°. And when we measure two parts of an angle and then the whole angle we know the closing true error because it should be 0. The difference here is called a discrepancy. It is a true error and not a residual error.

Example 176. If we measure a line n times under the same conditions the Median error of the average length is $\frac{r_0}{\sqrt{n}}$.

Hence if we measure it 4 times and the Median error of the average length is $\frac{r_0}{\sqrt{4}} = \cdot 03$, and $r_0 = \cdot 06$, and we want to get an average whose Median error will be less than ·001 it will be necessary to measure it n times, where $\frac{r_0}{\sqrt{n}} = \frac{\cdot 06}{\sqrt{n}} = \cdot 001$, or $n = 3600$ times. That is, we should use a more precise method. But if a Median error of the average length of ·01 will do we will have to measure it $n = 36$ times. The Median error of the average length for one trial is ·06. Three more trials reduce it to ·03. It takes 32 more trials to reduce it to ·01, and 3564 more trials to reduce it to ·001.

Thus it does not pay to measure anything more than a few times unless very great precision is required.

The Median error is normally in the middle, that is, if there are 17 errors there normally are 8 on each side of it, 8 smaller and 8 larger than it. The chance that the error will be on one side of it is $\frac{1}{2}$ and the chance that it will be on the other side of it is $\frac{1}{2}$.

Example 177. If the 100 foot tape we use in measuring a line is found to be 100·1 feet long, and we get 1000 feet for the length of the line, we must take $\frac{100\cdot 1}{100} \times 1000 = 1001$ feet for the length got. The excess in the length of the tape has no contradictory, for it is always the same, and hence does not follow the law of errors, by Axiom 3. It is sometimes called a constant error or expectation.

It saves a lot of trouble if the tape is of the precise length and all the instruments in adjustment.

Example 178. Since chance is unknown cause, by Definition 1, all known causes should be excluded as far as possible.

Any large defect not caused by a number of very small errors is a *mistake*. Mistakes follow the law of errors, but if they are very large the precision will be very small, since then the elements are not all very small.

Example 179. In fixing a point by a number of courses of distance and bearing the error due to bearing is proportional to the distance, and the error due to chainage is proportional to the square root of the distance

Hence when the courses are long the error due to chainage may be neglected, and the total error taken as the error due to bearings alone.

In country districts the total error should never be more than $\frac{1}{500}$ of the sum of the lengths of the courses, and in city surveying it should not exceed $\frac{1}{5000}$.

Example 180. Suppose now that the 4 *obs* in Example 169 were not made under the same conditions, but that their weights are 1, 2, 3 and 4, respectively.

The average length of the line is

$$\frac{p_1 l_1 + p_2 l_2 + p_3 l_3 + p_4 l_4}{p_1 + p_2 + p_3 + p_4}$$

$$= \frac{1000{\cdot}4 + 2000{\cdot}4 + 3001{\cdot}8 + 4003{\cdot}2}{1 + 2 + 3 + 4}$$

$= 1000{\cdot}58.$

$$\begin{aligned} v_1 &= -\,{\cdot}18 \\ v_2 &= -\,{\cdot}38 \\ v_3 &= +\,{\cdot}02 \\ v_4 &= +\,{\cdot}22 \\ v_1^2 &= {\cdot}0324 \\ v^2_2 &= {\cdot}1444 \\ v^2_3 &= {\cdot}0004 \\ v^2_4 &= {\cdot}0484. \end{aligned}$$

And to reduce these *obs* to weight 1 they must be taken a number of times equal to their respective weights, by Definition 13.

Hence for an *obs* of weight 1

$$p_1 v_1^2 = 1 \times \cdot 0324 = \cdot 0324.$$
$$p_2 v_2^2 = 2 \times \cdot 1444 = \cdot 2888.$$
$$p_3 v_3^2 = 3 \times \cdot 0004 = \cdot 0012.$$
$$p_4 v_4^2 = 4 \times \cdot 0484 = \cdot 1936.$$
$$\Sigma p v^2 = \cdot 5160.$$
$$\Sigma p = 1 + 2 + 3 + 4 = 10.$$
$$n - 1 = 3.$$

Hence the Normal error of an *obs* of weight 1 is

$$x_0 = \sqrt{\frac{\Sigma p v^2}{n-1}} = \sqrt{\frac{\cdot 5160}{3}} = \cdot 415.$$

And the Normal error of the average is

$$\sqrt{\frac{x_0^2}{\Sigma p}} = \sqrt{\frac{\Sigma p v^2}{(n-1)\,\Sigma p}} = \sqrt{\frac{\cdot 5160}{3 \times 10}} = \cdot 131.$$

And the Median error for an *obs* of weight 1 is $r = \cdot 280$.
And the Median error of the average is $r_0 = \cdot 088$.

Example 181. And the chance of an element c of error at $x = \cdot 5$ at the end of a single *obs* of weight 1 is

$$y = \frac{2c}{\sqrt{2\,\frac{\Sigma p v^2}{n-1}}}\, e^{-\frac{x^2}{2\,\frac{\Sigma p v^2}{n-1}}} = 1 \cdot 648c.$$

Example 182. What is the chance that the error will be greater than $\cdot 5$ at the end of a single *obs* of weight 1 ?

$$t = \frac{x}{\sqrt{2x_0^2}} = \frac{\cdot 5}{\sqrt{\cdot 344}} = \cdot 853.$$

Hence P = $\cdot 772$.
Hence 1 — P = $\cdot 228$, which is the answer.

Example 183. What is the chance that the total error will be greater than ·5 some time during any 3 *obs* each of weight 1 ?

$$y = e^{-\frac{x^2}{6x_0^2}} = e^{-\frac{x^2}{6\frac{\Sigma pv^2}{n-1}}} = \cdot 785.$$

Example 184. What is the chance that the error of the average will be greater than ·5 at the end of the series?

$$t = \frac{x}{\sqrt{2\frac{x_0^2}{\Sigma p}}} = \frac{x}{\sqrt{2\frac{\Sigma pv^2}{(n-1)\,\Sigma p}}} = \frac{\cdot 5}{\sqrt{\cdot 0344}} = 2\cdot 695.$$

Hence P = ·99986.

Hence 1 — P = ·00014, which is the answer.

Example 185. What is the chance that the error of the average will be greater than ·5 some time during the series ?

$$y = e^{-\frac{x^2}{2\frac{x_0^2}{\Sigma p}}} = e^{-\frac{\cdot 25}{\cdot 0344}} = \cdot 0005.$$

What is the chance that the error at the end of the series will be greater than $6x_a$?

It is $\frac{1}{3}\left(\frac{1}{2}\right)^{\frac{s^2}{2}} = \frac{1}{3}\left(\frac{1}{2}\right)^{18}$.

CHAPTER IV

INDIRECT OBSERVATIONS

Example 186. If we have the 3 equations

$$2l_1 + 3l_2 = 8$$
$$3l_1 + 2l_2 = 7$$
$$3l_1 - l_2 = 2$$

and the weight of a single *obs* is 1, what are the most likely values of l_1 and l_2?

These are called observation equations.

If there is no error in the first two *obs* equations $l_1 = 1$ and $l_2 = 2$.

But if these are the correct values the last *obs* equation gives $3\,l_1 - l_2 = 1$.

Hence there is an error in some or all of the equations, and we must find the most likely values of l_1 and l_2.

In order to find the most likely values of l_1 and l_2 we must get two independent equations, each including all the information, that is, including the whole three *obs* equations.

In order to get $2\,l_1$ something we will call L_1 was measured twice and $2l_1$ is the sum of these two measurements. Hence $2\,l_1$ is twice the average of these two measurements. And there are only half as many pairs of contradictories for one line as for two.

Hence the weight of l_1 in the first equation is twice the weight of a single *obs*, by Example 159.

But the weight of a single *obs* is 1.

ERRATA

Substitute for the last eight lines, page 100, and the first line, page 101 :—

The single *obs* in the first equation are $2l_1$ and $3l_1$.

And the weight of each is one.

And as $2l_1$ is twice as long as l_1 there are twice as many chances of a pair of contradictories for $2l_1$ as for a single *obs* of l_1.

And since the average is the same for both *obs* there are twice as many chances of a pair of contradictories for a single *obs* of l_1 as for the average of two single *obs* of l_1.

Hence the weight of l_1 in the first equation is $2 \times 2 = 2^2 = 4$, by Def. 13.

Substitute for the first thirteen lines, page 107 :—

A Normal equation is nothing but the reduction of the average of the parts of a single *obs* to weight 1, so as to compare the weights and find the values of different averages, which are the most likely values, by Prop. 12.

If a line of weight 1 is divided into n equal parts the weight of a single *obs* of one part is n, by Def. 13.

And if x_0 is the Normal error of a single *obs* of one part the Normal error for the whole line is $\sqrt{nx_0^2}$. And since the number of chances of error is divided into n equal parts for the average of the parts the Normal error of the average is $\frac{\sqrt{nx_0^2}}{n}$, by Prop. 11, and the weight of the average is n^2, by Def. 13.

In the first line, page 205, substitute x_a^2 is 4 for $2x_a^2$ is 2.

Hence the weight of l_1 in the first equation is 2.

Again if the elements are all very small the most likely error in any series is 0, by Proposition 12.

Hence if there is any number of unknown quantities the most likely value of any one is found by assuming that all the rest are correct, and that all the error is in this one.

Hence the most likely value of l_1 is found by assuming that l_2 is correct in the first equation.

And $2l_1 = 8 - 3l_2$.

And since the weight of l_1 is 2 we must take 2 l_1 instead of l_1 in order to reduce it to weight 1, since it is the average of 2 *obs.*

Hence if in the first equation the whole error is in l_1 we have

$$2(2l_1 = 8 - 3\ l_2) \text{ or } 4\ l_1 = 16 - 6\ l_2,$$

since we must not change the value of l_1, but take it twice as often.

Similarly we must take the second equation three times, and the third equation three times, to reduce them to weight 1.

Hence we have

$$4l_1 + 6l_2 = 16.$$
$$9l_1 + 6l_2 = 21.$$
$$9l_1 - 3l_2 = 6.$$

And their sum

$$22l_1 + 9l_2 = 43.$$

is called the *Normal equation for* l_1, in which the whole error is in l_1.

Similarly we will get the most likely value of l_2 by assuming that l_1 is correct in the first equation, and that all the error of the equation is in l_2.

And similarly for the other equations.

Hence we have

$$6l_1 + 9l_2 = 24.$$
$$6l_1 + 4l_2 = 14.$$
$$-3l_1 + l_2 = 2.$$

Since l_2 is minus in the *obs* equation we must multiply through by -1 to make it plus.

The sum of these three equations is $9l_1 + 14l_2 = 36$, which is the *Normal equation for* l_2, in which the whole error is in the positive value of l_2.

Hence we have the two Normal equations

$$22l_1 + 9l_2 = 43 = \text{A}.$$
$$9l_1 + 14l_2 = 36 = \text{B}.$$

from which to find l_1 and l_2.

The weights of A and B are each 1, since they are each *obs* once.

These two equations give

$$l_1 = \frac{14\text{A} - 9\text{B}}{308 - 81} = 1{\cdot}22 = \frac{14}{227}\text{A} - \frac{9}{227}\text{B}. \quad (1)$$

$$l_2 = \frac{22\text{B} - 9\text{A}}{308 - 81} = 1{\cdot}78 = \frac{22}{227}\text{B} - \frac{9}{227}\text{A}. \quad (2)$$

Hence these are the most likely values of l_1 and l_2.

$$\text{Hence A} = \frac{227}{14}l_1 + \frac{9}{14}\text{B}.$$

$$\text{B} = \frac{227}{22}l_2 + \frac{9}{22}\text{A}.$$

And the whole error in A is in l_1, and in B there is no error in l_1, since in B the whole error is in l_2.

And in measuring A, whose weight is 1, l_1 is measured $\frac{227}{14} = 16{\cdot}2$ times, and in measuring B, whose weight is 1, l_2 is measured $\frac{227}{22} = 10{\cdot}3$ times.

Hence the weight of l_1 is $p_1 = 16{\cdot}2$, and the weight of l_2 is $p_2 = 10{\cdot}3$, by Definition 13.

Hence in equations (1) and (2) the first term on the right of each equation is all that is required to find the weights, no matter how many terms there are, or how many Normal equations there are.

Hence for the purpose of finding the weight of l_1 all the Normal equations except the one in which the whole error is in l_1 may each be taken as 0 on the right, that is, B, C, D, etc., may each be taken as 0 for the purpose of finding the weight of l_1.

And similarly for the other weights.

Example 187. And in general if the *obs* equations are

$$a_1l_1 + b_1l_2 + c_1l_3 + \text{etc.} = n_1,$$
$$a_2l_1 + b_2l_2 + c_2l_3 + \text{etc.} = n_2,$$
$$a_3l_1 + b_3l_2 + c_3l_3 + \text{etc.} = n_3, \text{ and so on,}$$

the Normal equation for l_1 is

$$\Sigma a^2l_1 + \Sigma abl_2 + \Sigma acl_3 + \text{etc.} = \Sigma an = \text{A},$$

in which the whole error is in l_1.

And the Normal equation for l_2 is

$\Sigma abl_1 + \Sigma b^2l_2 + \Sigma bcl_3 + \text{etc.} = \Sigma bn = \text{B}$, in which the whole error is in l_2.

And the Normal equation for l_3 is

$\Sigma acl_1 + \Sigma bcl_2 + \Sigma c^2l_3 + \text{etc.} = \Sigma cn = \text{C}$, in which the whole error is in l_3, and so on.

There is the same number of Normal equations as of unknown quantities.

Hence the most likely values of all the unknown quantities, and their weights, can be found, as in Example 186.

Example 188. If the whole error in A, whose weight is 1, is in l_1, and l_1 is contained p_1 times in A, the number of chances of a pair of contradictories for A is p_1 times that for l_1, by Proposition 11.

Hence the weight of l_1 is p_1, by Definition 13.

Example 189. If there are n *obs* equations and q things observed what is the normal error for an *obs* of weight 1 ?

An *obs* equation of any number of things is a single *obs*, the same as an *obs* equation of one thing, for all *obs* are equations. If a tape is laid over a row of 6 oranges the weight of the *obs* is 1, and not 6, if the weight of a single *obs* is 1, but the weight of the *obs* of an orange is 6, since the total number of chances of a pair of contradictories is divided up into 6 parts, by Definition 13.

If there are n *obs* of the same thing the weight of the average *obs* is n if that of a single *obs* is 1, because it has been *obs* n times instead of once.

And the number of chances of a pair of contradictories for each average is $\frac{2\pi x_0^2}{n}$, where x_0 is the normal error for an *obs* of weight 1, by Proposition 11.

Hence the number of chances of a pair of contradictories for the q averages is $\frac{2\pi x_0^2}{n} \times q = \frac{2\pi q x_0^2}{n}$.

Hence $2\pi x_0^2 = 2\pi v_0^2 + \frac{2\pi q x_0^2}{n}$, by Definition 14.

$$x_0^2 = v_0^2 + \frac{q x_0^2}{n}.$$

$$\frac{x_0^2}{n} = \frac{v_0^2}{n - q}.$$

$$\frac{n x_0^2}{n} = \frac{n v_0^2}{n - q}.$$

$$\frac{\Sigma x^2}{n} = x_0^2 = \frac{\Sigma v^2}{n - q}.$$

Hence $x_0 = \sqrt{\dfrac{\Sigma v^2}{n - q}}$.

When $q = 1$ we have $x_0 = \sqrt{\dfrac{\Sigma v^2}{n - 1}}$, which is the same formula we got for n *obs* of the same thing.

Example 190. What is the normal error of an *obs* of weight 1 in Example 186 ?

We have for the most likely values of the equations

$$8 = 2(1{\cdot}22) + 3(1{\cdot}78)$$
$$7 = 3(1{\cdot}22) + 2(1{\cdot}78)$$
$$2 = 3(1{\cdot}22) - 1{\cdot}78.$$

Hence the residuals are

$$v_1 = 8 - 2{\cdot}44 - 5{\cdot}34 = + {\cdot}22$$
$$v_2 = 7 - 3{\cdot}66 - 3{\cdot}56 = - {\cdot}22$$
$$v_3 = 2 - 3{\cdot}66 + 1{\cdot}78 = + {\cdot}12, \text{ by Axiom 3.}$$
$$v_1^2 = {\cdot}0484$$
$$v_2^2 = {\cdot}0484$$
$$v_3^2 = {\cdot}0144$$
$$\Sigma v^2 = {\cdot}1112$$
$$n = 3$$
$$q - 2.$$

Hence $x_0 = \sqrt{\dfrac{\Sigma v^2}{n - q}} = \sqrt{\dfrac{{\cdot}1112}{3 - 2}} = {\cdot}333.$

Example 191. What is the Median error of an *obs* of weight 1 in Example 186 ?

$r = {\cdot}6745 x_0 = {\cdot}225.$

Example 192. What is the Median value of l_1 in Example 186 ?

Since the weight of a single *obs* is 1, and the weight of l_1 is 16·2, we have for the normal error of l_1, $p_1 x_0^2 = (\cdot 333)^2$, by Definition 13.

$$x_0^2 = \frac{(\cdot 333)^2}{p_1} = \frac{\cdot 1112}{16\cdot 2} = \cdot 0069.$$

Hence $x_0 = \cdot 08$.

And $r_0 = \cdot 6745 \times \cdot 08 = \cdot 05$.

Hence the Median value of l_1 is $1\cdot 22 \pm \cdot 05$.

Example 193. What is the Median value of l_2 in Example 186?

$$x_0 = \frac{\cdot 333}{\sqrt{p_2}} = \frac{\cdot 333}{\sqrt{10\cdot 3}} = \cdot 10.$$

And $r_0 = \cdot 6745 x_0 = \cdot 07$.

Hence the Median value of l_2 is $1\cdot 78 \pm \cdot 07$.

Example 194. What is the chance that the error in l_1 will be greater than ·2 at the end of the series of 3 *obs*?

$$t = \frac{x}{\sqrt{2x_0^2}} = \frac{\cdot 2}{\sqrt{2 \times \cdot 0069}} = \cdot 12$$

Hence $P = \cdot 13$.

Hence $1 - P = \cdot 87$, which is the answer.

Example 195. What is the chance that the error in l_1 will be greater than ·2 sometime during the three *obs*?

$$\text{It is } e^{-t^2} = e^{-\cdot 0144} = \frac{2 - \cdot 0144}{2 + \cdot 0144} = \cdot 99.$$

Example 196. If the weight of a single *obs* in the three *obs* equations in Example 186 is 1, 2 and 3, respectively, what is the normal error of an *obs* of weight 1?

We can reduce the single *obs* to weight 1 by taking it a number of times equal to its weight, by Definition 13.

The Normal error of an *obs* of weight 1 is $x_0 = \cdot409$. Hence the Normal error of an *obs* of weight 3 is

$$\frac{\cdot409}{\sqrt{3}} = \cdot236.$$

Example 198. What is the Normal error of l_1? The weight of l_1 is 45.

Hence the Normal error of l_1 is $\dfrac{\cdot409}{\sqrt{45}} = \cdot06$.

Example 199. What is the Median value of l_2? The weight of l_2 is 18·3.

Hence the Median error of l_2 is $\cdot6745\,\dfrac{\cdot409}{\sqrt{18\cdot3}} = \cdot065$.

Hence the Median value of l_2 is $1\cdot74 \pm \cdot07$.

Example 200. What is the chance that the error of l_2 will be more than ·3 at the end of the 3 *obs*?

The Normal error of l_2 is $x_0 = \cdot10$.

$$t = \frac{x}{x_0\sqrt{2}} = \frac{\cdot3}{\cdot1 \times 1\cdot41} = 2\cdot13.$$

Hence $1 - \mathrm{P} = \cdot0026$, which is the answer.

Example 201. What is the chance that the error of l_2 will be more than ·3 some time during the 3 *obs*?

It is $e^{-t^2} = e^{-4\cdot54} = \cdot01$.

Example 202. Suppose that there are some plums and oranges all mixed up and arranged in a number of rows, and a tape is laid along any row containing 2 plums and 3 oranges and the length is 8 inches, and the tape is now laid along another row containing 3 plums and 2 oranges and the length is 7 inches, and again the tape is laid along another row containing 3 plums, less an orange, that is, there are 3 plums in one row and an orange in the next row, and the length of the difference is 2 inches, what is the most likely length of a plum and an orange, respectively ?

We have the 3 *obs* equations :

$$2l_1 + 3l_2 = 8$$
$$3l_1 + 2l_2 = 7$$
$$3l_1 - l_2 = 2.$$

This is the same as Example 186.

Example 203. Chance is objective, for, data being given, our thinking will not change the chances. But other data will, by Definition 1. If we think that head is twice as likely as tail our thinking will not make it so. But our knowledge will, by Definition 1. If our thinking is wrong experience will show it in a very large number of trials, because error is proportional to only the square root of n, by Proposition 11. We can assume that head is twice as likely as tail if we like, and find the expectation on that assumption, but it may be checked by experience, which will pay no attention to our assumptions. If it did our inferences would be useless, and chance and error a plaything.

The chance that the error will be greater than sx_a at the end of any series is $1 - P = \frac{1}{3}\left(\frac{1}{2}\right)^{\frac{s^2}{2}}$, very nearly, for all values of s not less than 3, if the expectation for each trial is constant and all the elements are very small. And our thinking has nothing to do with it.

CHAPTER V

STATISTICS

Example 204. We have shown that when the number of pairs is normal $\sqrt{nx_0^2} = \sqrt{npq}$.

A report of the British Association gives the measurements of the heights of 6194 Englishmen taken by chance. The number from 57–58 inches in height was 1, from 58–59 inches was 3, and so on.

The following Table gives the results.

HEIGHT IN INCHES	NO. OF MEN
57	1
58	3
59	12
60	39
61	70
62	128
63	320
64	524
65	740
66	881
67	918
68	886
69	753
70	473
71	254
72	117
73	48
74	16
75	9
76	1
77	1

If the measurements are only given to differences of an inch we will suppose that the men between 67 and 68 inches are exactly 67.5 inches high, and so on.

The average height is about 67.5 inches, since the greatest number of men is about this height, and a small error is more likely than a larger one, by Proposition 12.

There are	881 men	1 inch less than this
	740	2
	524	3
	320	4
	128	5
	70	6
	39	7
	12	8
	3	9
	1	10

That is, altogether there are $881 \times 1 + 740 \times 2 + 524 \times 3 +$ and so on $= 6679$ inches less and $886 \times 1 + 753 \times 2 + 473 \times 3 +$ and so on $= 5903$ inches more than required to give each man 67.5 inches in height.

The difference 776 inches must be divided among the 6194 men, giving $67.5 - \cdot 125 = 67 \cdot 375 = 67\frac{3}{8}$ inches for the average height.

Now $6679 + 5903 = 12582$ would be the numerical sum of the residuals if 67.5 were the average height.

And if $67\frac{3}{8}$ is the average height this sum will be practically unchanged except that we must add $\frac{1}{8}$ inch for each man over 67·5 inches and subtract $\frac{1}{8}$ inch for each man under 67·5 inches in height.

Hence the sum will be $12582 + 95 = 12677$, approx.

Hence $\Sigma v = 12677$.

But we have shown that when n is very large

$$x_0 = 1{\cdot}2533 \frac{\Sigma v}{\sqrt{n(n-1)}} = \frac{5}{4} \cdot \frac{\Sigma v}{n}, \text{ very nearly.}$$

$$\text{And } r = {\cdot}8453 \frac{\Sigma v}{\sqrt{n(n-1)}} = \frac{11}{13} \cdot \frac{\Sigma v}{n}, \text{ very nearly.}$$

$$\text{Hence } x_0 = 1{\cdot}2533 \frac{12677}{\sqrt{6194 \times 6193}} = 2{\cdot}57 \text{ approx.}$$

And $r = {\cdot}6745x_0 = 1{\cdot}73$, approx.

Hence the Median height of an Englishman is $67{\cdot}375 \pm 1{\cdot}73$ inches, approx.

That is, it is an even bet that an Englishman will be between 69·11 and 65·65 inches in height.

The Median height of Englishmen is $67{\cdot}375 \pm \frac{1{\cdot}73}{\sqrt{n}}$ $= 67{\cdot}375 \pm {\cdot}015$.

Example 205. What is the chance that an Englishman will be at least six feet in height? And what is the chance of an error between $x - c$ and $x + c$ if c is very small?

For the first the error is $72 - 67{\cdot}375 = 4{\cdot}625$.

$$t = \frac{4{\cdot}625}{x_0\sqrt{2}} = \frac{4{\cdot}625}{3{\cdot}634} = 1{\cdot}27.$$

Hence $\frac{1 - P}{2} = {\cdot}036$, or 36 in 1000.

The formula for the Natural chance of an error of exactly x in any trial is

$$y = \frac{2}{\sqrt{2\pi x_0^2}} e^{-\frac{x^2}{2x_0^2}}, \text{ by Proposition 12 and Def. 12.}$$

We have shown that when n is very large

$$2x_0^2 = \pi \left(\frac{\Sigma x}{n}\right)^2.$$

$$\text{Hence } y = \frac{2}{\pi \frac{\Sigma x}{n}} e^{-\frac{x^2}{\pi\left(\frac{\Sigma x}{n}\right)^2}}.$$

And the chance of an error between $x - c$ and $x + c$ is

$$y = \frac{2 \times 2c}{\sqrt{2\pi x_0^2}} e^{-\frac{x^2}{2x_0^2}} = \frac{4c}{\pi \frac{\Sigma x}{n}} e^{-\frac{x^2}{\pi\left(\frac{\Sigma x}{n}\right)^2}}$$

if c is a very small element of x.

Example 206. What is the chance that an Englishman will be between 72.375 and 72·75 inches in height?

The errors are $72{\cdot}375 - 67{\cdot}375 = +5$ inches and $72{\cdot}75 - 67{\cdot}375 = +5{\cdot}375$ inches, or $x - \frac{c}{2}$ and $x + \frac{c}{2}$.

$$t_1 = \frac{5}{3{\cdot}634} = 1{\cdot}376, \text{ and } t_2 = \frac{5{\cdot}375}{3{\cdot}634} = 1{\cdot}479.$$

$$\text{Hence } \frac{P_1}{2} = \frac{{\cdot}94834}{2} = {\cdot}47417, \text{ and}$$

$$\frac{P_2}{2} = \frac{{\cdot}96353}{2} = {\cdot}48176.$$

Hence $\frac{P_2 - P_1}{2} = {\cdot}0076$, which is the chance that an Englishman will be between 72·375 and 72·75 inches in height.

Or since $\frac{5{\cdot}375 - 5}{2} = \frac{{\cdot}375}{2}$ is a very small element of $\frac{5{\cdot}375 + 5}{2} = 5{\cdot}1875$ the chance of an element c of error at x is

$$y = \frac{\cdot 375}{\sqrt{2\pi x_0^2}} e^{-\frac{x^2}{2x_0^2}} = \frac{\cdot 375}{2\cdot 57\sqrt{2\pi}} e^{-\frac{(5\cdot 1875)^2}{2(2\cdot 57)^2}}$$

$= \cdot 0076$ the same as before.

When $\frac{x^2}{\pi\left(\frac{\Sigma v}{n}\right)^2}$ is a very small fraction the expression

$$e^{-\frac{x^2}{\pi\left(\frac{\Sigma v}{n}\right)^2}} \text{ is very nearly equal to } \frac{2 - \frac{x^2}{\pi\left(\frac{\Sigma v}{n}\right)^2}}{2 + \frac{x^2}{\pi\left(\frac{\Sigma v}{n}\right)^2}}$$

$$= \frac{2\pi\left(\frac{\Sigma v}{n}\right)^2 - x^2}{2\pi\left(\frac{\Sigma v}{n}\right)^2 + x^2}, \text{ by Example 25.}$$

If $x = \frac{1}{2}$ this is equal to $\frac{25\cdot 72 - \cdot 25}{25\cdot 72 + \cdot 25}$

$= \cdot 98057$, since $\Sigma v = 12677$ and $n = 6194$.

And if x is next to 0 its value is 1.

Hence the Natural chance of an error of exactly next to 0 is only $\frac{1}{\cdot 98057} = 1\cdot 029$ times as great as that of an error of exactly $\frac{1}{2}$.

Since an error next to 0 only occurs $1\cdot 029$ times as often as an error of exactly $\frac{1}{2}$ there is very little error in taking the height of all the men between 67 and 68 inches as exactly $67\cdot 5$ inches in height, and giving each an error of $67\cdot 5 - 67\cdot 375 = +\frac{1}{8}$ inch.

And the error in taking the middle height for any group is small.

Example 207. What is the chance that an Englishman will be at least 7 feet in height?

The error is $84 - 67\frac{3}{8} = 16\frac{5}{8} = 16{\cdot}625$.

$$t = \frac{16{\cdot}625}{x_0\sqrt{2}} = \frac{16{\cdot}625}{3{\cdot}634} = 4{\cdot}58.$$

Hence $\frac{1 - \mathrm{P}}{2} = \frac{1 - {\cdot}999{,}999{,}999}{2} = {\cdot}000{,}000{,}000{,}05.$

Which is the chance that an Englishman will be at least 7 feet in height. He is eligible for the circus.

Example 208. What is the chance that an Englishman will be less than 5 feet in height?

The error is $60 - 67\frac{3}{8} = -7{\cdot}375$.

$$t = \frac{7{\cdot}375}{3{\cdot}634} = 2{\cdot}03.$$

Hence $\frac{1 - \mathrm{P}}{2} = {\cdot}002.$

That is, out of 1000 Englishmen the expectation is that there will be 2 less than 5 feet in height.

Example 209. What is the chance that an Englishman will be less than 55 inches in height?

The error is $55 - 67\frac{3}{8} = -12{\cdot}375$.

$$t = \frac{12{\cdot}375}{3{\cdot}634} = 3{\cdot}32.$$

Hence $\frac{1 - \mathrm{P}}{2} = {\cdot}000{,}000{,}7.$

That is, the expectation is that seven Englishmen out of any 10,000,000 are less than 55 inches in height.

Example 210. We can get the normal error by first finding the average height, and then the residual errors and then the sum of their squares.

We can also get it by first finding the average height, and then the residual errors, and then the sum of these all taken as plus.*

Example 211. If the sum of the heights of 1000 Scotsmen and 10,000 Englishmen in a district is 62,000 feet, and in another district the sum of the heights of 8,000 Scotsmen and 1000 Englishmen is 51,000 feet, and in another district that of 3,000 Englishmen and 2,000 Scotsmen is 28,000 feet, and the weights of the *obs* in the three different districts are 1, 2 and 3, respectively, what are the most likely heights of an Englishman and Scotsman, respectively?

And what is the chance that an Englishman will be over 6 feet in height? What is the chance that a Scotsman will be over 6 feet high? What is the Median height of Englishmen and of Scotsmen, respectively?

This is the same as Example 196, and those following it

Example 212. If $\frac{1}{10}$ of the soldiers in the world war are killed, and 4,000 go from a place, what is the chance that more than 500 of them will be killed?

The error is 100.

$$t = \frac{x}{\sqrt{2npq}} = \frac{100}{\sqrt{2 \times 4000 \times \frac{1}{10} \times \frac{9}{10}}} = 3{\cdot}73.$$

Hence $\frac{1 - \text{P}}{2} = {\cdot}000{,}000{,}06.$

Example 213. What is the chance that more than 450 will be killed?

$t = 1{\cdot}86.$

Hence $\frac{1 - \text{P}}{2} = {\cdot}0004.$

* See Yule's *Theory of Statistics.*

Example 214. What is the chance that more than 400 will be killed?

It is $\frac{1}{2}$.

Example 215. What is the chance that less than 300 will be killed?

It is the same as the chance that more than 500 will be killed.

Example 216. What is the Median error?

$$r = \cdot 6745\sqrt{npq} = 13.$$

Hence it is an even chance that between $np \pm 13 = 400 \pm 13$ or between 387 and 413 will be killed.

Example 217. What is the chance that a particular man will be killed?

It is $\frac{1}{10} = \cdot 1$.

Example 218. What is the chance that two particular men will be killed?

It is $\left(\frac{1}{10}\right)^2 = \frac{1}{100} = \cdot 01.$

Example 219. If three men, A B and C, go to the war, what is the chance that they will all be killed?

It is $\left(\frac{1}{10}\right)^3 = \frac{1}{1000} = \cdot 001.$

Example 220. What is the chance that two of them will be killed and the other not killed?

It is $\frac{\lfloor 3}{\lfloor 2 \, \lfloor 1}\left(\frac{1}{10}\right)^2 \times \frac{9}{10} = \frac{27}{1000} = \cdot 027.$

Example 221. What is the chance that one will be killed and the other two not killed ?

$$\text{It is } \frac{|\underline{3}}{|\underline{1}\;|\underline{2}} \times \frac{1}{10} \times \left(\frac{9}{10}\right)^2 = \frac{243}{1000} = \cdot 243.$$

Example 222. What is the chance that none of them will be killed ?

$$\text{It is } \left(\frac{9}{10}\right)^3 = \frac{729}{1000} = \cdot 729.$$

Example 223. We can get all these results from the formula

$$(p+q)^3 = p^3 + \frac{|\underline{3}}{|\underline{2}\;|\underline{1}} p^2 q + \frac{|\underline{3}}{|\underline{1}\;|\underline{2}} p q^2 + q^3$$

$$= \left(\frac{1}{10}\right)^3 + 3\left(\frac{1}{10}\right)^2\left(\frac{9}{10}\right) + 3\left(\frac{1}{10}\right)\left(\frac{9}{10}\right)^2 + \left(\frac{9}{10}\right)^3.$$

The first term is the chance that all will be killed. The second that two will be killed and the other not killed, and so on.

The sum of all these chances is 1.

Example 224. What is the chance that some of them will be killed ?

$$\text{It is } 1 - \left(\frac{9}{10}\right)^3 = \cdot 271.$$

Example 225. What is the chance that some of them will not be killed ?

$$\text{It is } 1 - \left(\frac{1}{10}\right)^3 = \cdot 999.$$

Example 226. What is the chance that at least two of them will be killed ?

It is $\left(\frac{1}{10}\right)^3 + \frac{\lfloor 3}{\lfloor 2 \, \lfloor 1} \left(\frac{1}{10}\right)^2 \left(\frac{9}{10}\right) = \cdot 028.$

Example 227. The chances p and q, as in a coin, have nothing to do with chance, but with the shape of the coin. It is s in Prop. 12 that is ruled by chance.

If we know the chances they are both subjective and objective, because knowledge is the conformity of the subjective with the objective.

The chance that A committed a certain crime, that anyone might have committed cannot be calculated precisely on any evidence, unless it must have been committed by one or more of a given set of persons, for we may not know all the persons relevant to the case. Or even if we did. See Keynes' *Probability*.

Example 228. If the expectation is that the U-boats will sink 20 ships a week, and the normal error is 5, what is the chance that in any week they will sink more than 30 ships ?

The error is 30 — 20 = 10.

$$t = \frac{10}{x_0 \sqrt{2}} = 1{\cdot}41.$$

Hence $\frac{1 - P}{2} = \cdot 02.$

Hence the expectation is that they will sink more than 30 ships in any week about once a year.

Example 229. In progressive cards, if the chance to win a set at any table is $\frac{1}{2}$ and there are four sets played, what is the chance that a person will win 0, 1, 2, 3 or 4 sets, respectively ?

The chances are

$(p+q)^4 = p^4 + 4p^3q + 6p^2q^2 + 4pq^3 + q^4.$

The chance to win 0 sets is $(\frac{1}{2})^4 = \frac{1}{16}$.

The chance to win 1 set is $4\ (\frac{1}{2})^1\ (\frac{1}{2})^3 = \frac{1}{4}$

The chance to win 2 sets is $6(\frac{1}{2})^4 = \frac{3}{8}$.

The chance to win 3 sets is $4(\frac{1}{2})^4 = \frac{1}{4}$.

The chance to win 4 sets is $(\frac{1}{2})^4 = \frac{1}{16}$.

Thus if the players are all equally good a person is as likely to win all the sets as he is to win none, and as likely to win 3 sets as he is to win 1.

He is more likely to win 3 sets than he is to win none. He is 6 times as likely to win 2 sets as he is to win none. And he is 4 times as likely to win 1 set as he is to win none.

If, however, he has more or less skill than others p and q will be different for him, and the chances will be very different for a good player from those for an indifferent player.

A good player will win 3 or 4 sets more often.

And in statistics of these things a very precise value of the chances can be found, and whether they are constant or variable, as in Chapter IX. But things with very different expectations should not be put in the same aggregate, or if they are they should be analysed separately, for expectations do not follow the law of error unless the differences are very small, by Proposition 11. *Anything* in a manageable general formula for error when n is very large must have a specific *nature*, that is, a constant *a priori* expectation for each trial, as Proposition 6, because the error is measured from the expectation, by Definition 11. Index numbers are variable expectations.

Chapter VI

TARGET PRACTICE

Example 230. If a soldier shoots 17 shots at a target the median error is normally the middle error. Hence a simple way for him to find his median error at Target Practice is to shoot an odd number, say 17, shots at a target and measure the distance from the centre of the target to the centre of the 9th best shot.

If this distance is 1 inch and the range is 25 yards his median error is about 1 inch at 25 yards, or 4 inches at 100 yards. That is, at 100 yards he would put half his shots in a bull's-eye of 8 inches diameter, and at 600 yards he would put half his shots in a bull's-eye of 4 feet diameter.

If he repeats this a few times and takes the average the result will not differ much from his median error at 25 yards.

We should, however, for great precision take the square root of the average of the squares of the error, by Proposition 11.

But the difference is very slight if there is a large number of shots.

Thus if one has a chance to examine the card targets that have been used by the individuals of a company of soldiers he will be able to compare their skill very quickly and precisely by finding their median errors from their cards.

Since the chance of an error greater than r is

$y = e^{-\frac{r^2}{r_0^2}}$, if we put $e^{-\frac{r^2}{r_0^2}} = \frac{1}{2}$

$\log\left(e^{-\frac{r^2}{r_0^2}}\right) = \log \frac{1}{2}$.

Hence $\frac{r^2}{r_0^2} = \log_e 2$, and $r^2 = r_0^2 \log_e 2$.

If we put a for this value of r we have

$a^2 = r_0^2 \log_e 2$, where a is the median error, since the chance of an error greater than it is $\frac{1}{2}$.

We have shown that $2^x = e^{x \log_e 2}$.

Hence $2^{\frac{x}{\log_e 2}} = e^x$.

Hence $e^{-\frac{r^2}{r_0^2}} = 2^{-\frac{r^2}{r_0^2 \log_e 2}} = 2^{-\frac{r^2}{a^2}} = (\frac{1}{2})^{\frac{r^2}{a^2}}$.

Since $(\frac{1}{2})^{\frac{r^2}{a^2}}$ is much simpler to use than $e^{-\frac{r^2}{r_0^2}}$ it is better to use $(\frac{1}{2})^{\left(\frac{r}{a}\right)^2}$ than $e^{-\frac{r^2}{r_0^2}}$ for the chance of an error greater than r.

And $1 - (\frac{1}{2})^{\left(\frac{r}{a}\right)^2}$ for the chance of an error less than r.

Example 231. If a soldier's median error at 25 yards is $a = 1$ inch, what is the chance that he will hit a bull's-eye 6 inches in diameter at 25 yards?

$$1-\left(\tfrac{1}{2}\right)^{\left(\frac{r}{a}\right)^2}=1-\left(\tfrac{1}{2}\right)^9=\frac{511}{512}.$$

Hence the chance that he will hit the bull's-eye is $\frac{511}{512}$. That is, out of 512 shots it is his expectation to make 511 bull's-eyes.

Example 232. What is the chance that he will hit a ring between 2 and 3 inches from the centre ?

The chance that he will hit in a circle 2 inches in radius is $1-\left(\tfrac{1}{2}\right)^4=\frac{15}{16}$.

Hence the chance that he will hit the ring between 2 and 3 inches radii is $\frac{511}{512}-\frac{15}{16}=\frac{31}{512}$.

That is, if he shoots 512 shots his expectation is to put 31 of them in this ring.

Example 233. What is the chance that he will put from 28 to 34 in this ring in a series of 512 shots ?

$$p=\frac{31}{512},\ q=\frac{481}{512}.$$

$$t=\frac{x}{\sqrt{2npq}}=\frac{3}{\sqrt{2\times 512\times\frac{31}{512}\times\frac{481}{512}}}=\cdot 39.$$

Hence P = ·42, which is the chance that he will put from 28 to 34 in the ring.

Example 234. What is the Median error of the number of shots in this ring ?

$$r=\cdot 6745\ \sqrt{npq}=3\cdot 65.$$

That is, it is an even chance that he will put from 27.35 to 34.65 shots in the ring.

Example 235. What is the Median error of his median error ?

The expectation of his median error is to put half the shots in his median circle, and he is equally likely to put more or less than this in it.

Hence $p = q = \frac{1}{2}$.

Hence $r = \cdot 6745 \sqrt{512 \times \frac{1}{2} \times \frac{1}{2}} = 7 \cdot 6$.

Hence it is an even chance that he will put from $256 - 7\cdot 6 = 248\cdot 4$ to $256 + 7\cdot 6 = 263\cdot 4$ shots in a bull's-eye of radius 1 inch or diameter 2 inches out of 512 shots at a range of 25 yards.

Example 236. What is the chance that out of 512 shots he will put less than 240 shots in this bull's-eye ?

The error must be greater than -16.

$$t = \frac{x}{\sqrt{2npq}} = \frac{16}{\sqrt{2 \times 512 \times \frac{1}{2} \times \frac{1}{2}}} = 1.$$

Hence $P = \cdot 8427$.

Hence $\frac{1 - P}{2} = \cdot 0786$, which is the chance that the error will be more than -16.

Hence the chance that out of 512 shots he will put less than 240 shots in a bull's-eye of 2 inches diameter at a range of 25 yards if his median error at 25 yards is 1 inch is ·0786, or once in about 13 different series of 512 shots each.

Example 237. What is the chance that he will miss a bull's-eye of radius 3 inches at a range of 200 yards ?

His median error at 100 yards is $a = 1 \times \frac{100}{25} = 4$ ins.

And for l hundred yards is $al = 4l$ inches.

$$y = (\tfrac{1}{2})^{\left(\frac{r}{al}\right)^2} = (\tfrac{1}{2})^{\frac{9}{64}}.$$

We have shown that when the elements are all very small

$$(\tfrac{1}{2})^{\frac{9}{64}} = \frac{3 - \frac{9}{64}}{3 + \frac{9}{64}} = \frac{183}{201}.$$

Hence his expectation is to miss it 183 times out of 201 shots.

Example 238. What is the chance that he will miss a bull's-eye of 3 feet diameter at a range of 200 yards ?

$$y = (\tfrac{1}{2})^{\left(\frac{r}{al}\right)^2} = (\tfrac{1}{2})^{\left(\frac{18}{4 \times 2}\right)} = (\tfrac{1}{2})^{\frac{81}{16}} = (\tfrac{1}{2})^{5\frac{1}{16}}$$

$$= (\tfrac{1}{2})^{5} \times (\tfrac{1}{2})^{\frac{1}{16}}$$

$$= \frac{1}{32} \times \frac{3 - \frac{1}{16}}{3 + \frac{1}{16}} = \frac{1}{32} \times \frac{47}{49} = \frac{47}{1568}.$$

That is, his expectation is to miss the bull's-eye 47 times out of 1568 shots.

Example 239. What is the chance that he would miss a target of 15 inches radius at 200 yards ?

$$y = (\tfrac{1}{2})^{\left(\frac{r}{al}\right)^2} = (\tfrac{1}{2})^{\left(\frac{15}{4 \times 2}\right)^2} = (\tfrac{1}{2})^{3\frac{33}{64}}$$

$$= (\tfrac{1}{2})^{4-\frac{31}{64}} = (\tfrac{1}{2})^{4} \times (\tfrac{1}{2})^{-\frac{31}{64}}$$

$$= \frac{1}{16} \times \frac{3+\frac{31}{64}}{3-\frac{31}{64}} = \frac{1}{16} \times \frac{223}{161} = \cdot 0866, \text{ approx.}$$

That is, his expectation is to miss it 866 times out of 10,000 shots.

Example 240. We have shown that the Natural chance of an error of r is

$$y = \frac{2\pi r}{\pi r_0^2} e^{-\frac{r^2}{r_0^2}}. \quad \text{See Example 22.}$$

Now if for any value of r an element of y divided by an element of r is 0 the above expression must be either a maximum or a minimum, because it is neither getting greater nor less as r changes.

When $r = 0$, $y = 0$.

Hence the minimum value of y is 0.

The ratio between an element of y and an element of

r is $\left(\frac{2}{r_0^2} - \frac{4r^2}{r_0^4}\right) e^{-\frac{r^2}{r_0^2}} = 0$ for a maximum.

Hence $r = \frac{r_0}{\sqrt{2}} = \frac{x_0\sqrt{2}}{\sqrt{2}} = x_0 = \cdot 8495a$ for a maximum value of y.

Hence the most likely error is $r = \cdot 8495a$, or about $\frac{11}{13}$ of the median error.

Example 241. Hence if $al = 8$ inches there will be

more shots strike the circumference of a circle of radius $\frac{11}{13}$ of $8 = 7$ inches than any other circumference.

And $\frac{3}{4}$ of the shots, which is called the *effective fire*, extends to $r = a\sqrt{2} = \frac{5}{3}x_0$.

The most likely error is about $\frac{11}{13}$ of the median error for any range. The greatest number of shots strike just inside the median circle.

Example 242. If out of 500 shots 260 fall in a circle of radius 1 inch, what is the median error ?

The chance of missing a circle of r inches radius is

$$y = (\tfrac{1}{2})^{\left(\frac{r}{a}\right)^2}.$$

$$\text{Hence } (\tfrac{1}{2})^{\left(\frac{r}{a}\right)^2} = \frac{500 - 260}{500} = \frac{12}{25}.$$

$$\text{Hence } \left(\frac{r}{a}\right)^2 \log 2 = \log 25 - \log 12.$$

$$\text{Hence } a = r\sqrt{\frac{\log 2}{\log 25 - \log 12}}$$

$$= 1 \times \sqrt{\frac{\cdot 30103}{\cdot 31876}} = \cdot 972 \text{ inches.}$$

Example 243. Thus if we visited a rifle club and examined the card targets that had been used by the members, and counted 260 shots inside a circle of 1 inch radius, and knew that the total number of shots fired was 500, and that the range was 25 yards, we would know their skill very closely.

Their median error for a range of 100 yards is $\cdot 972 \times 4 = 3\cdot 89$ inches. They would put half their shots in a circle of 7·78 inches diameter at a range of 100 yards.

Thus it is easy to compare the skill of two or more rifle clubs or individuals.

It is well to take a circle about the size of the median circle so as not to isolate the stragglers.

Example 244. We have shown that the Natural chance of an error of r in target practice is

$$y = \frac{2\pi r}{\pi r_0^{\,2}} e^{-\frac{r^2}{r_0^{\,2}}} = \frac{2r}{r_0^{\,2}} \left(\tfrac{1}{2}\right)^{\frac{r^2}{r_0^{\,2}\log_e 2}} .$$

And that $r_0^{\,2} = \dfrac{a^2}{\log_e 2}$.

$$\text{Hence } y = \frac{2 \times \cdot 69315\, r}{a^2} \left(\tfrac{1}{2}\right)^{\frac{r^2}{a^2}}$$

$$= \frac{1\cdot 3863\, r}{a^2} \left(\tfrac{1}{2}\right)^{\frac{r^2}{a^2}} .$$

Example 245. Hence if the thickness of a ring is $2c$, and c is a very small element of r, the chance of hitting the ring is

$$y = \frac{1\cdot 3863\, r}{a^2} \left(\tfrac{1}{2}\right)^{\left(\frac{r}{a}\right)^2} \times 2c$$

$$= \frac{11cr}{(2a)^2} \left(\tfrac{1}{2}\right)^{\left(\frac{r}{a}\right)^2}, \text{ approx.}$$

Or if a is the median error for 100 yards and l is the number of hundred yards of range

$$y = \frac{11cr}{(2al)^2} \left(\tfrac{1}{2}\right)^{\left(\frac{r}{al}\right)^2}.$$

Example 246. If $l = \frac{1}{4}$, $a = \frac{1}{3}$ feet, $c = \cdot 01$ feet and $r = \cdot 15$ feet, what is the chance of hitting the ring whose middle is r from the centre of the target ?

$$y = \frac{11 \times \cdot 01 \times \cdot 15}{(2 \times \frac{1}{3} \times \frac{1}{4})^2} \left(\tfrac{1}{2}\right)^{\left(\frac{\cdot 15}{\frac{1}{3} \times \frac{1}{4}}\right)^2}$$

$$= \cdot 594 \left(\tfrac{1}{2}\right)^{3 \cdot 24}$$

$$= \cdot 594 \times \tfrac{1}{8} \times \frac{3 - \cdot 24}{3 + \cdot 24} = \frac{276}{324} = \cdot 063, \text{ approx.}$$

That is, the expectation is that 63 shots out of 1000 will be put in the ring at the range of 25 yards.

Example 247. We have shown that the Natural chance of hitting a point r from the centre is

$$y = \frac{1}{\pi r_0^{\,2}} e^{-\frac{r^2}{r_0^{\,2}}} = \frac{\cdot 69315}{\pi a^2} \left(\tfrac{1}{2}\right)^{\left(\frac{r}{a}\right)^2}.$$

Or if a is the median error at 100 yards and l is the number of 100 yards of range

$$y = \frac{2}{(3al)^2} \left(\tfrac{1}{2}\right)^{\left(\frac{r}{al}\right)^2}, \text{ approx.}$$

This may be taken as the density of shots at any point r from the centre of the target.

Example 248. If $a = \frac{1}{3}$ feet, $l = 6$ and $r = 0$ the density at the centre of the target is

$$\frac{2}{(3 \times \frac{1}{3} \times 6)^2} = \frac{1}{18}.$$

If $r = 2$ feet the density at any point 2 feet from the centre of the target is

$$\frac{1}{18}\left(\tfrac{1}{2}\right)^{\left(\frac{2}{\frac{1}{3} \times 6}\right)^2} = \frac{1}{18} \times \tfrac{1}{2}.$$

Thus at a range of 600 yards if a soldier's median error for 100 yards is $\frac{1}{3}$ feet any point 2 feet from the centre of the target would be $\frac{1}{2}$ as dangerous as at the centre.

Example 249. The danger at a point 1 foot from the centre is

$$\left(\tfrac{1}{2}\right)^{\frac{1}{4}} = \frac{3 - \frac{1}{4}}{3 + \frac{1}{4}} = \frac{11}{13} \text{ of that at the centre.}$$

Example 250. At 6 feet from the centre the danger is

$$\left(\tfrac{1}{2}\right)^{\left(\frac{6}{\frac{1}{3} \times 6}\right)^2} = \left(\tfrac{1}{2}\right)^9 = \frac{1}{512} \text{ as much as at the centre.}$$

Example 251. At a distance of 10 feet from the centre the danger is only $\left(\tfrac{1}{2}\right)^{\left(\frac{10}{\frac{1}{3} \times 6}\right)^2} = \left(\tfrac{1}{2}\right)^{25}$

$= \cdot 000{,}000{,}03$ as much as at the centre.

Example 252. The error is measured from the expectation. If the gun is properly sighted for the range, wind, and so on, the expectation is 0. But when the range is long, gravity causes a large expectation, which is known and allowed for in sighting.

If the target is in motion the point of aim must be advanced to the amount of the expectation. In shooting at a wild duck it is usual to aim 2 feet ahead of it, and 6 times the length of a flying machine.

But this depends on the range. At long range the expectation of the wind is often several feet.

Example 253. The Natural chance of an error of x at the end of a series of n trials in measuring a line or in games of chance is $\frac{2}{\sqrt{2\pi\Sigma x^2}} e^{-\frac{x^2}{2\Sigma x^2}}$.

Hence the ratio between the Natural chance of an error of x some time during a series and the Natural chance of an error of x at the end of the series in measuring a line or in games of chance is

$$\frac{\frac{2\pi x}{2\pi\Sigma x^2}}{\frac{2}{\sqrt{2\pi\Sigma x^2}}} = \frac{x}{\sqrt{\Sigma x^2}}\sqrt{\frac{\pi}{2}} = \frac{x}{\sqrt{npq}}\sqrt{\frac{\pi}{2}}.$$

This ratio is k when $x = k\sqrt{npq}\sqrt{\frac{2}{\pi}} = k\sqrt{\Sigma x^2}\sqrt{\frac{2}{\pi}}$ $= \cdot 8k\sqrt{npq} = \cdot 8k\sqrt{\Sigma x^2}$, very nearly, when the elements are all very small, by Definition **14**.

And $k = \frac{x}{x_0\sqrt{n}}\sqrt{\frac{\pi}{2}} = \frac{x}{x_a\sqrt{n}}$, by Proposition **11**.

Example 254. In a series of 100 flips of a coin the Natural chance of an error of exactly x some time during the series is k times the Natural chance of an error of exactly x at the end of the series when

$$x = \cdot 8k\sqrt{100 \times \tfrac{1}{2} \times \tfrac{1}{2}} = 4k.$$

Example 255. What is the value of x when the chance of an error of exactly x some time during a series of 10,000 flips of a coin is the same as the chance of an error of exactly x at the end of the series ?

The value of x is 40, or the average error at the end.

Example 256. When $x = 0$ what is the ratio between the chance of an error of x some time during a series and an error of x at the end of the series ?

$k = 0$.

Example 257. If $k = 2$, $x_0 = \cdot 25$, and $x = 60$ what is the value of n ?

$$\sqrt{n} = \frac{x}{\cdot 8kx_0} = 150.$$

Hence $n = 22500$.

Example 258. What is the Natural chance that there will be an error of exactly 25 some time during 1000 flips of a coin ?

It is the same as the Natural chance of a total error of $x = 25$ in 1000 shots at a target, or

$$y = \frac{2\pi x}{2\pi\Sigma x^2} e^{-\frac{x^2}{2\Sigma x^2}} = \frac{x}{nqp} e^{-\frac{x^2}{2npq}}$$

$$= \frac{25}{1000 \times \frac{1}{2} \times \frac{1}{2}} e^{-\frac{(25)^2}{2 \times 1000 \times \frac{1}{2} \times \frac{1}{2}}} = \frac{1}{10} e^{-1\cdot 25}$$

$$= \frac{1}{10} \times \cdot 2865 = \cdot 0287.$$

Example 259. What is the most likely error some time during a series of 1000 flips of a coin ?

We have shown that the most likely error in target practice is $x_0 \sqrt{n} = \sqrt{1000 \times \frac{1}{2} \times \frac{1}{2}} = 15{\cdot}81$, which is the most likely error some time during a series of 1000 flips of a coin.

Example 260. There is one very marked difference between the chance of an error of exactly x at the end of a series of n trials and the chance of an error of exactly x some time during a series of n trials, which is that the chance of a small error at the end of the series is always greater than the chance of a larger one, and the most likely error is 0. But the most likely error some time during a series of n trials is not 0, but x_0, or the Normal error, or $x_0 \sqrt{n}$ when the total error is taken.

Example 261. In measuring a line all the errors congregate around 0, but in shooting at a target they do not, but around the ring x_0 from the centre. It is very exasperating when shooting at a target to put so many shots near this ring, which is not far from the centre, and yet to be unable to get many much closer to the centre.

Example 262. It is the same with the error some time during a series of n trials at any game or measurement. The errors do not congregate around the point of no error but around the error x_0, or Normal error, or $x_0 \sqrt{n}$ when the total error is taken. The most likely error is x_0 and not 0.

Example 263. Since in the case of the error at the end of the series the most likely error is 0 the Natural chance formulæ applied to Common chance problems give the best result when x is near 0, because the elements of error are smaller there. But in the case of the error some time during the series the most likely error is $\sqrt{npq}$, and the Natural chance formulæ applied to Common chance problems give the best result when x is near this value, because the elements of error are smaller there, if x is not too small in comparison with 1.

In Proposition I suppose the n things to be n years each divided up into 12 months. If an earthquake is equally likely to occur in any of the total number of months, but cannot occur twice in any month, it may occur 12 times in any year. That is, any of the things may be taken 12 times. Thus we can make a combination containing 12 things from a single thing when it is divided up into 12 elements. But after it has occurred in any month in any year it is only $\frac{11}{12}$ as likely to occur again in that year, by Def. 3.

And if a thing is divided up into an infinite number of elements the thing may be taken an infinite number of times, and a combination containing an infinite number of things can be made from one thing. And they are all equally likely to be taken at any time. And all given combinations are equally likely. And the sum of the chances of a combination from n things is c^n. But if things can be repeated, as in Example 61, the sum of the chances of a combination is infinite.

CHAPTER VII

ERRORS IN THREE DIMENSIONS

Example 264. To hit a point in the surface of a sphere of radius r we must make the errors x_1, near or far, x_2 in a direction at right angles to this direction and x_3 in a direction at right angles to each of the other directions, that is, we have to make the error $r = \sqrt{x_1^2 + x_2^2 + x_3^2}$, by Axiom 7.

These three directions are at right angles to each other, for if they are not, two or all may have a common element, by Axiom 6.

And they all occur at the same time.

Hence the Natural chance of an error r in any given direction is $\left(\frac{1}{\sqrt{2\pi x_0^2}}\right)^3 e^{-\frac{r^2}{2x_0^2}}$, by Propositions 3 and 11 and Definitions 12 and 19 and Axioms 5 and 6, since errors are either wins or failures.

And the point may be in any part of the shell of radius r, surface $4\pi r^2$ and thickness 1, where 1 is a very small element of r, by Definition 14 and 19. See Example 22.

Hence the Natural chance of an error r is

$$\frac{4\pi r^2}{\left(\sqrt{2\pi x_0^2}\right)^3} e^{-\frac{r^2}{2x_0^2}} = \frac{r^2}{2x_0^2} \cdot \frac{4}{\sqrt{2\pi x_0^2}} e^{-\frac{r^2}{2x_0^2}},$$

where x_0 is the Normal error in each of the three directions at right angles to each other.

Example 265. Hence the most likely error is

$$r = x_0\sqrt{2}.$$

Hence in shooting at an aeroplane more shells will burst on the surface of a sphere of radius $r = x_0\sqrt{2}$ from the centre of the machine than on the surface of any other sphere if the Normal error is the same for each of the three directions and the expectation is taken correctly.

Example 266. And the chance of bursting a shell between $r - c$ and $r + c$ from the centre of the flying machine is $y = \frac{r^2}{2(lx_0)^2}\cdot\frac{4 \times 2c}{\sqrt{2\pi}\,(lx_0)^2}\, e^{-\frac{r^2}{2(lx_0)^2}}$,

where l is the number of 100 yards of range, and x_0 is the Normal error in each of the three directions at right angles to one another for 100 yards,

$$= \frac{4cr^2}{(lx_0)^3\sqrt{2\pi}}\, e^{-\frac{r^2}{2(lx_0)^2}} = 1\cdot 6\,\frac{cr^2}{(lx_0)^3}\, e^{-\frac{r^2}{2(lx_0)^2}},$$ very

nearly, if c is a small element of r.

Example 267. Hence an element of the sum of the chances of making an error less than r is

$$\frac{4\pi r^2 c}{(\sqrt{2\pi x_0^2})^3}\, e^{-\frac{r^2}{2x_0^2}}$$

$$= \frac{4\pi r^2 c}{2\pi x_0^2\sqrt{2\pi x_0^2}}\, e^{-\frac{r^2}{2x_0^2}}$$

$$= \frac{4}{\sqrt{\pi}} \cdot \frac{r^2}{2x_0^{\,2}}\, e^{-\frac{r^2}{2x_0^{\,2}}} \left(\frac{c}{\sqrt{2x_0^{\,2}}}\right)$$

$$= \frac{4}{\sqrt{\pi}}\, t^2 e^{-t^2}\, dt, \text{ where } \frac{c}{\sqrt{2x_0^{\,2}}} = dt = k \text{ is an element}$$

of t,

$$= \frac{4k}{\sqrt{\pi}}\, t^2 \left(1 - t^2 + \frac{t^4}{\underline{|2}} - \frac{t^6}{\underline{|3}} + \frac{t^8}{\underline{|4}} - \text{etc.}\right)$$

$$= \frac{4k}{\sqrt{\pi}} \left(t^2 - t^4 + \frac{t^6}{\underline{|2}} - \frac{t^8}{\underline{|3}} + \frac{t^{10}}{\underline{|4}} - \text{etc.}\right).$$

Hence the sum of the chances is the integral

$$\frac{4}{\sqrt{\pi}} \left(\frac{t^3}{3} - \frac{t^5}{5} + \frac{t^7}{7\underline{|2}} - \frac{t^9}{9\underline{|3}} + \frac{t^{11}}{11\underline{|4}} - \text{etc.}\right), \text{ which}$$

is the chance that the error will be less than r.

Example 268. It will be seen that when t is about $1{\cdot}1$ this is equal to $\frac{1}{2}$.

Hence when $t = 1{\cdot}1 = \frac{r}{\sqrt{2x_0^{\,2}}}$ half the shells will burst within a sphere of radius $r = x_0 \times 1{\cdot}1\sqrt{2} = 1{\cdot}5x_0$, approx., whose centre is the centre of the target.

Example 269. Hence the median error is $1{\cdot}5x_0$, approx.

If $x_0 = 1$ foot and $l = 100$, half the shells will burst within a radius of 150 feet from the centre of the target, and if $l = 50$ half the shells will burst within a radius of 75 feet from the centre of the target, if the expectation is taken correctly.

Example 270. There will be more shells burst on the surface of a sphere of radius $lx_0\sqrt{2} = 1{\cdot}41\,lx_0$ than on the surface of a sphere of any other radius, that is, the most likely error is about $\frac{14}{15}$ of the median error.

Example 271. Hence if there are 100 shells fired at a flying machine at a range of a mile there should be about 50 burst within 75 feet of its centre.

Example 272. The Natural chance of hitting a point on the surface of a sphere of radius r from the centre of the target is

$$y = \frac{1}{\left(\sqrt{2\pi (lx_0)^2}\right)^3} e^{-\frac{r^2}{2(lx_0)^2}}.$$

If $r = 0$ and $lx_0 = 50$ this is $y = \cdot 000{,}000{,}5$, which may be called the danger at the centre of the target.

Example 273. At a distance of 10 feet from the centre the danger is

$$e^{-\frac{r^2}{2(lx_0)^2}} = e^{-\cdot 02} = \frac{2 - \cdot 02}{2 + \cdot 02} = \frac{99}{101}$$

as much as at the centre, if $x_0 = 1$ foot and $l = 50$.

Example 274. At a distance of 75 feet the danger is $e^{-1\cdot 125} = \cdot 28$ as much as at the centre.

Example 275. At a distance of 100 feet from the centre the danger is $e^{-\frac{10,000}{5,000}} = e^{-2} = \cdot 135$ as much as at the centre.

Example 276. At a distance of 200 feet from the centre the danger is $e^{-8} = \cdot 0003$ as much as at the centre.

Example 277. At 300 feet it is ·000,000,02 as much.

Chapter VIII

MONTE CARLO

Example 278. If A's and B's chances to win any game are ·49 and ·51, respectively, what is the chance that at the end of 100 games A will be ruined and B not ruined if A has $1 and B has $2, and they each stake 10 cents a game ?

B's expectation for 100 games is 100 (·02 × ·1) = ·2. And A's is — ·2.

Hence an error of ·8 at the end of the series will ruin A, and an error of 2·2 will ruin B.

The normal error in games is $\sqrt{nx_0^2} = \sqrt{\cdot 49 \times \cdot 51 n}$ $= x_0 \sqrt{n} = 5$, by Definition 14.

The chance that the sum of the stakes will tend to be both won and lost at the same time is ·49 × ·51 = $\frac{1}{4}$, very nearly.

The sum of the stakes is ·2.

Hence the normal cash error is

$$\sqrt{nx_0^2} = \sqrt{\cdot 49 \times \cdot 51 \times (\cdot 2)^2 n} = \cdot 1 \sqrt{n} = 1$$

$$\sqrt{2nx_0^2} = 1\cdot 41.$$

For A,

$$t = \frac{x}{\sqrt{2nx_0^2}} = \frac{\cdot 8}{1\cdot 41} = \cdot 57.$$

Hence P = ·58, which is the chance that the error will be less than ·8.

Hence the chance that the error will be more than ·8 is 1 — P = ·42.

Hence the chance that the loss will be more than ·8 is $\frac{1-P}{2} = \cdot 21$, by Example 152.

Which is the chance that A will be ruined at the end of the series.

For B,

$$t = \frac{x}{\sqrt{2nx_0^2}} = \frac{2\cdot 2}{1\cdot 41} = 1\cdot 56.$$

Hence P = ·97.

Hence $\frac{1-P}{2} = \cdot 02$.

Which is the chance that B will be ruined at the end of the series.

Hence the chance that B will not be ruined is 1 — ·02 = ·98.

Hence the chance that A will be ruined and B not ruined is ·21 × ·98 = ·21.

Example 279. What is the chance that B wlll ruin A some time during the series ?

The normal total error for *obs* is $x_0 \sqrt{n}$.

The most likely total error in Target practice is $x_0 \sqrt{n}$.

Hence the most likely error some time during the series at any game is also $x_0 \sqrt{n} = 1$, by Proposition 12

Hence most of the errors will be near 1.

Since A's expectation changes at each game it will take a larger error to ruin him early in the series than later.

But since the variation in his fortune from 1 to ·8 is not very great we may take the average, or ·9, as the error that will ruin him. But the precision would be greater for the same range if it were farther from the most likely error.

Hence the chance that he will be ruined some time during the series is $\frac{1}{2}e^{-\frac{x^2}{2nx_0^2}} = \frac{1}{2}e^{-\frac{\cdot 81}{2}} = \cdot 334$, by Definition 14.

The chance that B will be ruined some time during the series is $\frac{1}{2}e^{-\frac{(2\cdot 1)^2}{2}} = \cdot 055$, by Definition 14.

Hence the chance that A will be ruined and B not ruined some time during the series is $(1 - \cdot 055) \times \cdot 334 = \cdot 316$.

And if both would be ruined some time during the series, but A ruined first, B would ruin A.

A is $k = \dfrac{\cdot 334}{\cdot 055} = 6$ times as likely to be ruined some time during the series as B is.

If they were equally likely to be ruined some time during the series, and both were ruined, the chance that A would be ruined first is $\dfrac{k}{k+1} = \dfrac{1}{1+1} = \frac{1}{2}$, which is exact for $k = 1$.

And it cannot be far wrong for any value of k.

The chance that they would both be ruined some time during the series is $\cdot 334 \times \cdot 055 = \cdot 018$, by Proposition 3 and Definition 6.

Hence the chance that B will ruin A is $\cdot 316 + \dfrac{k}{k+1}$ of $\cdot 018 = \cdot 33$, approx.

Example 280. If at any kinds of games a player has a chance of $\frac{1}{3}$ to win 6, a chance of $\frac{1}{6}$ to win 12, a chance of $\frac{3}{4}$ to lose 4 and a chance of $\frac{1}{4}$ to lose 4·04, and all these chances are independent, what is the chance that he will win 10 or more at the end of 100 games of each ?

He has a chance of $\frac{1}{3}$ to win 6.

Hence he has a chance of $1 - \frac{1}{3} = \frac{2}{3}$ not to win 6, by Definition 3.

Hence the normal number of pairs of game contradictories in n games is $2\pi n \times \frac{1}{3} \times \frac{2}{3}$, by Proposition 11.

And this is the normal number of pairs of contradictories, by Definition 8.

But for convenience we may multiply these by the square of the sum of the stakes instead of dividing the error by the sum of the stakes, since the result is the same, and call the product cash contradictories. See Example 127.

Hence the normal number of pairs of cash contradictories is $2\pi n \times \frac{1}{3} \times \frac{2}{3} \times 6^2$.

Similarly for the second chance the normal number of pairs of cash contraditories is $2\pi n \times \frac{1}{6} \times \frac{5}{6} \times 12^2 = 2\pi n \times 20$.

And for the third chance is $2\pi n \times \frac{1}{4} \times \frac{3}{4} \times 4^2 = 2\pi n \times 3$.

And for the fourth chance is $2\pi n \times \frac{1}{4} \times \frac{3}{4} \times (4{\cdot}04)^2 = 2\pi n \times 3{\cdot}063$.

Each of these is precise, but their sum is only approximate unless the sum of the stakes is the same in each case.

Hence the total number of chances of a pair of cash contradictories is $2\pi n\,(8 + 20 + 3 + 3{\cdot}063) = 214n$, approx.

The normal cash error for n games is

$$x_0\sqrt{n} = \sqrt{n}\,\sqrt{8 + 20 + 3 + 3{\cdot}063} = 5{\cdot}836\,\sqrt{n} = 58{\cdot}36$$

The Median cash error is $r = \cdot 67448 \times 58\cdot 36\sqrt{n} = 39\cdot 36$.

The expectation for n games is

$\frac{1}{3}$ of 6 + $\frac{1}{6}$ of 12 — $\frac{3}{4}$ of 4 — $\frac{1}{4}$ of 4·04 = — ·01n = — 1.

He must gain more than $10 + 1 = 11 = x$.

$$t = \frac{x}{\sqrt{2nx_0^2}} = \frac{11}{82\cdot 54} = \cdot 133.$$

Hence P = ·149.

$$\frac{1 - \text{P}}{2} = \cdot 43, \text{ which is the answer.}$$

There must be a large number of games for great precision.

Example 281. He has a chance of
$\frac{1}{4}$ to win more than 38·36.
$\frac{1}{4}$ to win less than 38·36.
$\frac{1}{4}$ to lose more than 40·36.
$\frac{1}{4}$ to lose less than 40·36.

Example 282. What is the chance that he will lose more than 10 at the end of 100 games?

He must lose more than $10 - 1 = 9 = x$.

$$t = \frac{x}{\sqrt{2nx_0^2}} = \frac{9}{82\cdot 54} = \cdot 109.$$

Hence $\frac{\text{I} - \text{P}}{2} = \cdot 44$.

Example 283. What is the chance that he will lose more than 10 some time during 100 games?

Most of the errors are near $x_0\sqrt{n} = 58$.

We will take the average fortune 9·5, by Definition **14.**

$x = 9\cdot 5$.

$$t^2 = \left(\frac{9{\cdot}5}{82{\cdot}54}\right)^2 = {\cdot}013.$$

$$p = \tfrac{1}{2}e^{-t^2} = \tfrac{1}{2}\left(\frac{2 - {\cdot}013}{2 + {\cdot}013}\right) = {\cdot}49.$$

When the limiting error is much greater than the most likely error there is not much difference between the chance of an error greater than the former some time during and at the end of the series. See Example **517**.

Example 284. If there are 6 men, of whom A is one, playing Pool, and each stakes 1 each game, and A's chance to win any game is $\frac{1}{4}$, what is the chance that A will win more than 100 at the end of 100 games ?

His expectation is n ($\frac{1}{4}$ of 5 — $\frac{3}{4}$ of 1) $= \frac{n}{2} = 50$.

His normal number of pairs of game contradictories is $2\pi n\ (\frac{1}{4} \times \frac{3}{4})$.

The sum of the stakes is 6.

The normal number of pairs of cash contradictories is $2\pi n\ (\frac{1}{4} \times \frac{3}{4}) \times 6^2 = 2\pi n{x_0}^2$, where x_0 is the normal cash error for 1 game.

$$2n{x_0}^2 = 2n(\tfrac{1}{4} \times \tfrac{3}{4}) \times 6^2 = 2n \times \frac{27}{4} = 1350.$$

$$\sqrt{2n{x_0}^2} = 36{\cdot}74.$$

His expectation wins him 50.

Hence he must gain more than $100 - 50 = 50 = x$.

$$t = \frac{x}{\sqrt{2n{x_0}^2}} = \frac{50}{36{\cdot}74} = 1{\cdot}36.$$

$$\frac{1 - \text{P}}{2} = {\cdot}027 \text{ which is the answer.}$$

Example 285. What is the chance that he will lose more than 100 at the end of 100 games?

$x = 100 + 50 = 150.$

$t = \dfrac{150}{36 \cdot 74} = 4 \cdot 1.$

$\mathrm{P} = \cdot 999{,}999{,}992.$

$\dfrac{1 - \mathrm{P}}{2} = \cdot 000{,}000{,}004$, which is the answer.

Example 286. What is the chance that he will lose more than 100 some time during 100 games?

Most of the errors are near $x_0\sqrt{n} = 26$.

We will take the average fortune, by Definition 14.

His expectation at the 50th game is 25.

$x = 100 + 25 = 125.$

$t^2 = \left(\dfrac{x}{\sqrt{2nx_0{}^2}}\right)^2 = \left(\dfrac{125}{36 \cdot 74}\right)^2 = (3 \cdot 4)^2 = 11 \cdot 56.$

$p = \frac{1}{2}e^{-11 \cdot 56} = \cdot 000{,}009$, by Definition 14, which is the answer.

Example 287. At Monte Carlo there is a roulette, or little wheel, and the man who owns it is called B or the Bank.

The chance that the Bank will be ruined is very small.

Hence we may neglect it, by Definition 14.

There are 37 numbers from 0 to 36 on the wheel. The 0 is not coloured, and 18 of the other numbers are red and 18 black.

When any money is staked it no longer belongs to the one who staked it, but must be won by some one or more. If A stakes a on red or black he has the same chance to win the stakes $2a$ that the bank has, with the exception that if 0 turns up the bank wins $\frac{3}{2}a$ and A only $\frac{1}{2}a$ from the stakes $2a$.

Hence when 0 turns up A loses half his stake, or $\frac{1}{2}a$.

The chance that 0 will turn up is $\frac{1}{37}$.

Hence A's expectation in a series of n games is

$$-\frac{1}{37} \text{ of } \frac{na}{2} = -\frac{na}{74} = -\cdot 0135na, \text{ by Definition 11.}$$

The chance that A will win the stakes $2a$ is $\frac{18}{37}$, and the chance that he will not is $\frac{19}{37}$.

Hence his normal number of pairs of cash contradictories is $2\pi n \left\{\frac{18}{37} \times \frac{19}{37} \times (2a)^2\right\} = 2\pi n \times \cdot 99927a^2$
$= 2\pi n x_0^2$.

Hence the normal error is $x_0\sqrt{n} = a\sqrt{n}$, by Definition 14. And this is the most likely total error in Target practice.

Example 288. If A plays 900 games and stakes 1 each game, what is his Median cash error for the series ?

It is $r\sqrt{n} = \cdot 6745x_0\sqrt{n} = 20$.

Example 289. What is his Median error in games ?
The sum of the stakes is 2.

Hence his Median error in games is $\frac{20}{2} = 10$.

Example 290. What is his Median gain and loss ?

His expectation is $-\frac{900}{74} = -12$.

Hence his Median gain is $20 - 12 = 8$, and his Median loss is $20 + 12 = 32$.

He has a chance of

$\frac{1}{4}$ to win more than 8;
$\frac{1}{4}$ to win less than 8;
$\frac{1}{4}$ to lose more than 32;
$\frac{1}{4}$ to lose less than 32.

Example 291. What is the chance that at the end of 10,000 games he will win more than 100?

In 10,000 games the adverse expectation is 135. Hence if there is no error he will lose 135, by Definition 11. Hence in order to win more than 100 he must gain more than $100 + 135 = 235$.

$$t = \frac{x}{\sqrt{2nx_0^2}} = \frac{235}{141} = 1{\cdot}67.$$

Hence P = ·98, which is the chance that the error will be less than 235.

Hence the chance that the error will be more than 235 is $1 - P = {\cdot}02$.

Hence the chance that the error will be more than $+ 235$ is $\frac{1 - P}{2} = .01$, by Example 152.

Which is the chance that he will win more than 100 in 10,000 games.

Example 292. What is the chance that he will lose more than 100 at the end of 10,000 games?

His expectation loses him 135.

Hence he will lose more than 100 unless he gains more than $135 - 100 = 35 = x$.

$$t = \frac{x}{\sqrt{2nx_0^2}} = \frac{35}{141} = \cdot 25.$$

Hence P = ·276, which is the chance that the error will be less than 35.

Hence the chance that the error will be more than 35 is 1 — P = ·724.

Hence the chance that the gain will be more than 35 is $\frac{1 - P}{2} = \cdot 362$, by Example 152.

Hence the chance that the gain will not be more than 35 is $1 - \frac{1 - P}{2} = \frac{1 + P}{2} = \cdot 64$.

Which is the chance that he will lose more than 100 at the end of the series.

Example 293. What is the chance that he will lose more than 100 some time during 10,000 games ?

His expectation at the end of the series is — 135.

Most of the errors are near $x_0\sqrt{n} = 100$.

We cannot take the average fortune.

If there is no error his fortune is lost in the 7407th game.

The chance that he will be ruined some time during 7407 games cannot be less than the chance that he will be ruined at the end of 7407 games, which is ½.

Hence for the first 7407 games we must take his fortune as 0, since $\frac{1}{2}e^{\frac{0^2}{2nx^2}} = \frac{1}{2}$.

Hence the result ·64 in Example 292 is very close for this case, since the limiting error is farther from the most likely error than 0 is.

Example 294. If A plays on a single number the Bank stakes 35 to his 1.

Hence if he stakes a the Bank stakes $35a$.

Hence the sum of the stakes is

$$a + b = a + 35a = 36a.$$

0 is an ordinary number now.

A's chance to win is $\frac{1}{37}$ and his chance to fail is $\frac{36}{37}$ by Definition 3.

Hence his expectation for 1 game is

$$d_1 = \frac{1}{37} \text{ of } 35a - \frac{36}{37} \text{ of } a = -\frac{a}{37} = -\cdot 02703a.$$

The normal number of pairs of game contradictories is

$$2\pi npq = 2\pi n\left(\frac{1}{37} \times \frac{36}{37}\right) = 2\pi n \times \cdot 02630.$$

The normal number of pairs of cash contradictories is $2\pi npq\,(a + b)^2$.

For cash $x_0 = 5{\cdot}83784$.

$\sqrt{2nx_0{}^2} = 8{\cdot}256\sqrt{n}.$

The Median error in games for the whole series is

$$r\sqrt{n} = \cdot 67448x_0\sqrt{n} = \cdot 10938\sqrt{n}.$$

The Median cash error is $\cdot 10938\,(a + b)\sqrt{n} = 3{\cdot}93768a\sqrt{n}$.

The error is measured from the expectation.

He has a chance of

$\frac{1}{4}$ to win more than $a\,(3{\cdot}938\sqrt{n} - \cdot 027n)$.

$\frac{1}{4}$ to win less than $a\,(3{\cdot}938\sqrt{n} - \cdot 027n)$.

$\frac{1}{4}$ to lose more than $a\,(3{\cdot}938\sqrt{n} + \cdot 027n)$.

$\frac{1}{4}$ to lose less than $a\,(3{\cdot}938\sqrt{n} + \cdot 027n)$

at the end of the series.

When n is small there is not much difference between the gain and loss, but when n becomes large there is a great difference.

If a man plays only a few games he has a fair chance, but if he plays a great many games he has practically no chance at all.

Example 295. If A stakes 1 each game, what is the chance that he will lose more than 100 at the end of 10,000 games ?

His expectation is — 270.

Hence unless he gains more than 170 he will lose more than 100.

$$t = \frac{170}{8{\cdot}256\sqrt{10{,}000}} = {\cdot}206.$$

$$P = {\cdot}23.$$

$$\text{Hence } \frac{1+P}{2} = {\cdot}62.$$

Which is the answer.

Example 296. What is the chance that he will win more than 100 ?

$$t = \frac{370}{8{\cdot}256\sqrt{10{,}000}} = {\cdot}448.$$

$$\frac{1-P}{2} = {\cdot}26.$$

Example 297. What is the chance that he will lose 1000 or more some time during a series of 10,000 games ?

The errors congregate around the circle $x_0\sqrt{n} = 584$ from the centre.

We will take the average fortune 865, by Definition 14.

$$\tfrac{1}{2}e^{-\frac{(865)^2}{681607}} = \cdot 45.$$

When the limiting error is very near to the most likely error the range through which the expectation may vary is very small for a precise result. When they are very far apart the range is large for the same precision.

Since the expectation is proportional to n and the normal number of combinations of error to $\sqrt{n}$ the fewer games A plays the better chance he has.

Example 298. $e^{-2\cdot341} = e^{-2}e^{-\cdot341} = e^{-2}\left(\frac{2 - \cdot341}{2 + \cdot341}\right)$, by Example 25.

As $\log e = \cdot4343$ it is very easy to find the value of e^{-2} mentally.

And $e^{-2\cdot873} = e^{-3}e^{+\cdot127} = e^{-3}\left(\frac{2 + \cdot127}{2 - \cdot127}\right)$.

CHAPTER IX

VARIABLE CHANCES

Example 299. The chance of $n - r$ wins and r failures in n trials is $\frac{n!}{(n-r)!\,r!} p^{n-r} q^r$, by Proposition 5.

And this is true whether p and q are constant or variable, by Proposition 3.

Example 300. What is the chance of drawing 2 kings from a pack of 52 cards in 4 draws ?

$p_1 = \frac{4}{52}$, to draw a king.

$p_2 = \frac{3}{51}$, to draw another king.

$q_3 = \frac{48}{50}$, to draw a not king.

$q_4 = \frac{47}{49}$, to draw another not king.

Hence $p^{n-r} q^r = p^2 q^2 = p_1 p_2 q_3 q_4$ where p and q are variable.

And $\frac{n!}{(n-r)!\,r!} = 6.$

Hence the chance is $\frac{6 \cdot 4 \cdot 3 \cdot 48 \cdot 47}{52 \cdot 51 \cdot 50 \cdot 49}$.

Example 301. If there are m coins of which a are dimes, what is the chance of drawing $s - r$ dimes and r not dimes in s draws ?

It is $\dfrac{\underline{|s}}{\underline{|s-r}\,\underline{|r}}\,p^{s-r}\,q^{r}$.

$p_1 = \dfrac{a}{m}$, to draw a dime.

$p_2 = \dfrac{a-1}{m-1}$, to draw another dime.

$p_{s-r} = \dfrac{a-\overline{s-r-1}}{m-\overline{s-r-1}}$, to draw the s-rth dime.

$q_{s-r+1} = \dfrac{m-a}{m-\overline{s-r}}$, to draw a not dime.

$q_{s-r+2} = \dfrac{m-a-1}{m-\overline{s-r}-1}$ to draw another not dime.

$q_{s-r+3} = \dfrac{m-a-2}{m-\overline{s-r}-2}$, to draw another not dime.

$q_{s} = \dfrac{m-a-\overline{r-1}}{m-\overline{s-1}}$, to draw the rth not dime.

Hence $p^{s-r}\,q^{r} = p_1 p_2 p_3 \cdots p_{s-r}\,q_{s-r+1}\,q_{s-r+2} \cdots\cdots q_{s}$.

Hence the chance is

$$\frac{\underline{|s}}{\underline{|s-r}\,\underline{|r}} \cdot \frac{\underline{|a}}{\underline{|a-\overline{s-r}}} \cdot \frac{\underline{|m-\overline{s-r}}}{\underline{|m}} \cdot \frac{\underline{|m-a}}{\underline{|m-a-r}}.$$

$$\frac{\underline{|m-s}}{\underline{|m-s-r}} = \frac{\underline{|s}}{\underline{|s-r}\,\underline{|r}} \cdot \frac{\underline{|a}}{\underline{|a-s-r}} \cdot \frac{\underline{|m-s}}{\underline{|m}} \cdot \frac{\underline{|m-a}}{\underline{|m-a-r}}.$$

Example 302. If there are 10 coins, of which 4 are dimes, what is the chance of drawing 4 dimes in 4 draws ?

Example 303. If there are 10 coins of which 4 are dimes, what is the chance of drawing 3 dimes and 5 not dimes in 8 draws ?

It is $\frac{\underline{|8}}{\underline{|3}\,\underline{|5}} \cdot \frac{\underline{|4}}{\underline{|1}} \cdot \frac{\underline{|6}}{\underline{|10}} \cdot \frac{\underline{|2}}{\underline{|1}} = \frac{24}{45} = \cdot 5333.$

Example 304. If there are 10 coins, of which 3 are dimes, what is the chance of not drawing any dimes in 6 draws ?

$s - r = 0.$

Hence $s = r$.

Hence the chance is

$$\frac{\underline{|m-a}}{\underline{|m}} \cdot \frac{\underline{|m-s}}{\underline{|m-a-s}} = \frac{\underline{|7}}{\underline{|10}} \cdot \frac{\underline{|4}}{\underline{|1}} = \frac{1}{30} = \cdot 0333.$$

Example 305. In how many ways can 3 dimes and 5 not dimes be drawn from 10 coins, of which 4 are dimes in 8 draws ?

A different combination of 3 dimes can be made from 4 dimes in $\frac{\underline{|4}}{\underline{|3}\,\underline{|1}} = 4$ ways.

And a different combination of 5 not dimes can be made from 6 not dimes in $\frac{\underline{|6}}{\underline{|5}\,\underline{|1}} = 6$ ways.

Hence a different combination of 3 dimes and 5 not dimes can be made from 10 coins in $4 \times 6 = 24$ ways.

Or a different combination of 8 coins can be made from 10 coins in $\frac{\underline{|10}}{\underline{|8}\ \underline{|2}} = 45$ ways.

And the chance that a combination will be 3 dimes and 5 not dimes is $\frac{8}{15}$.

Hence a combination of 3 dimes and 5 not dimes can be made in $\frac{8}{15} \times 45 = 24$ ways, by Definition 11.

Example 306. If there are 3 horses in a stable and the chance that any horse will be taken is $\frac{1}{4}$, what is the chance that none of them will be taken?

It is $(1 - p)^3 = (\frac{3}{4})^3 = \frac{27}{64}$.

$$= 1 - 3p + 3p^2 - p^3, \text{ by Proposition 2.}$$

$$= 1 - 3 \times \tfrac{1}{4} + 3 \times (\tfrac{1}{4})^2 - (\tfrac{1}{4})^3 = \frac{27}{64}.$$

And in general

$$(1 - p)^s = 1 - sp + \frac{s(s-1)}{\underline{|2}} p^2 - \frac{s(s-1)(s-2)}{\underline{|3}}$$

p^3 + etc., by Proposition 2.

Example 307. The latter method is often much more convenient. In fact, when p is variable the former method cannot be used.

If the chance that any horse will be taken is p_1 and the chance that any 2 horses will be taken is p_2 and the chance that they will all be taken is p_3, the chance that none will be taken is

$$1 - 3p_1 + 3p_2 - p_3.$$

We cannot work this by the former method unless

$$p_2 = p_1^2 \text{ and } p_3 = p_1^3.$$

Example 308. If there are 3 horses in a stable and the chance that any horse will be taken is $\frac{1}{4}$, and the chance that any 2 horses will be taken is $\frac{1}{8}$, and the chance that they will all be taken is $\frac{1}{32}$, the chance that none will be taken is $1 - 3\cdot\frac{1}{4} + 3\cdot\frac{1}{8} - \frac{1}{32} = \frac{19}{32}$.

The varying chance here depends on the preference of the owner for 1, 2 or 3 horses. And p_2 does not equal $p^2{}_1$ and p_3 does not equal p_1^3.

But since $(1 - p)^3 = \frac{19}{32}$

$1 - p = \left(\frac{19}{32}\right)^{\frac{1}{3}}$ for the whole, but not for the parts.

Example 309. The number of ways in which n things can be given to s persons is s^n.

The first thing can be given to any of the s persons.

Hence the first thing can be given in s ways.

Similarly each thing can be given in s ways.

Hence the n things can be given in s^n ways, by Axiom 8.

Example 310. The number of ways in which n things can be given to s persons is s^n, by Example 309.

The number of ways in which n things can be given to $s - 1$ persons is $(s - 1)^n$, and to $s - 2$ persons is $(s - 2)^n$ and to $s - r$ persons is $(s - r)^n$.

Hence if n things are given to s persons the chance that any given r persons will get nothing is

$$\frac{(s - r)^n}{s^n} = \left(\frac{s - r}{s}\right)^n.$$

Hence the chance that any

$$\text{person will get 0 is } p_1 = \left(\frac{s-1}{s}\right)^n$$

$$\text{2 persons will get 0 is } p_2 = \left(\frac{s-2}{s}\right)^n$$

$$r \text{ persons will get 0 is } p_r = \left(\frac{s-r}{s}\right)^n.$$

Hence the chance that each will get some is

$$1 - s\left(\frac{s-1}{s}\right)^n + \frac{s(s-1)}{\lfloor 2}\left(\frac{s-2}{s}\right)^n -$$

$$\frac{s(s-1)(s-2)}{\lfloor 3}\left(\frac{s-3}{s}\right)^n + \text{etc.}$$

$$= \Sigma \frac{\lfloor s}{\lfloor s-r \, \lfloor r} \cdot \left(\frac{s-r}{s}\right)^n \left(-1\right)^r, \text{ by Example } \mathbf{306}.$$

Example 311. If 4 oranges are given by chance to 3 boys, what is the chance that each will get some ?

It is $\Sigma \dfrac{\lfloor 3}{\lfloor 3-r \, \lfloor r} \left(\dfrac{3-r}{3}\right)^4 \left(-1\right)^r$, in which r successively takes the values 0, 1, 2 and 3.

Hence the chance is

$$(1-p)^3 = 1 - 3\left(\tfrac{2}{3}\right)^4 + 3\left(\tfrac{1}{3}\right)^4 = \tfrac{4}{9} = \cdot 4444.$$

Example 312. In how many ways can 4 oranges be given to 3 boys so that each will get some ?

The chance that each will get some is $\frac{4}{9}$.

And they can be given in 3^4 ways.

Hence they can be given so that each will get some in $\frac{4}{9}$ of $3^4 = 36$ ways, by Definition 3.

Or in $3^4 \left\{1 - 3\left(\tfrac{2}{3}\right)^4 + 3\left(\tfrac{1}{3}\right)^4\right\}$.

$= 3^4 - 3 \cdot 2^4 + 3 \cdot 1^4 = 36$ ways.

Or the 4 oranges can be divided up into the combinations

4 0 0 and arranged in $\dfrac{4!}{4!\,0!\,0!\,2!} \times 3! = 3$

3 1 0 and arranged in $\dfrac{4!}{3!\,1!\,0!} \times 3! = 24$

2 2 0 and arranged in $\dfrac{4!}{2!\,2!\,0!\,2!} \times 3! = 18$

2 1 1 and arranged in $\dfrac{4!}{2!\,1!\,1!\,2!} \times 3! = 36$ ways,

by Example 37 and Axioms 9 and 8.

Hence they can be given in

$3 + 24 + 18 + 36 = 81 = 3^4$ ways.

All get some in 36 ways.

1 gets 0 in $24 + 18 = 42$ ways.

2 get 0 in 3 ways.

And in general n things can be given to s persons so that each will get some in

$$s^n - \frac{s!}{(s-1)!\,1!}(s-1)^n + \frac{s!}{(s-2)!\,2!}(s-2)^n$$

$$- \frac{s!}{(s-3)!\,3!}(s-3)^n + \text{etc.}$$

$$= \Sigma \frac{s!}{(s-r)!\,r!}(s-r)^n(-1)^r \text{ ways.}$$

CHAPTER X

SORTS

Example 313. In how many ways can 4 horses be put into 3 stables?

In $3^4 = 81$ ways, by Example 309.

Example 314. In how many ways can 4 horses be put into 3 stables so as to leave none of the stables empty?

In $3^4 - 3 \cdot 2^4 + 3 \cdot 1^4 = 36$ ways.

Example 315. In how many ways can a combination of 3 horses be made from 4 stables each containing a single horse?

The number of combinations of 4 things 3 at a time is $\frac{\underline{|4}}{\underline{|3}\ \underline{|1}} = \frac{\underline{|n}}{\underline{|s}\ \underline{|n-s}} = 4$, by Proposition 2, which is the answer.

Example 316. In how many ways can 3 like horses be put into 4 stables if only one horse can be put into any stable?

In $\frac{\underline{|n}}{\underline{|s}\ \underline{|n-s}} = \frac{\underline{|4}}{\underline{|3}\ \underline{|1}} = 4$ ways, by Definition 4 and 6, and Example 315.

Example 317. In how many ways can a combination of 4 horses be made from 3 stables each containing at least 4 horses?

In $\dfrac{\lfloor\!\underline{4+3-1}}{\lfloor\!\underline{4}\;\lfloor\!\underline{3-1}} = \dfrac{\lfloor\!\underline{n+s-1}}{\lfloor\!\underline{n}\;\lfloor\!\underline{s-1}} = \dfrac{3\cdot4\cdot5\cdot6}{\lfloor\!\underline{4}} = 15,$

by Example 61.

Example 318. In how many ways can a group of 4 horses be made from 3 stables?

A combination of 4 horses can be made in

$$\frac{\lfloor\!\underline{4+3-1}}{\lfloor\!\underline{4}\;\lfloor\!\underline{3-1}} \text{ ways.}$$

And each combination of 4 horses has $\lfloor\!\underline{4}$ groups, by Axiom 9.

Hence a group of 4 horses can be made in

$$\frac{\lfloor\!\underline{4+3-1}}{\lfloor\!\underline{4}\;\lfloor\!\underline{3-1}} \times \lfloor\!\underline{4} = \frac{\lfloor\!\underline{4+3-1}}{\lfloor\!\underline{3-1}} = 360 \text{ ways.}$$

Similarly a group of n things can be made from s kinds of things in $\dfrac{\lfloor\!\underline{n+s-1}}{\lfloor\!\underline{s-1}}$ ways.

Example 319. In how many ways can 4 horses be arranged in 3 stables?

They can be returned to the 3 stables in the same orders in which they can be taken out.

Hence 4 horses can be arranged in 3 stables in

$$\frac{\lfloor\!\underline{4+3-1}}{\lfloor\!\underline{3-1}} = \frac{\lfloor\!\underline{n+s-1}}{\lfloor\!\underline{s-1}} = 360 \text{ ways.}$$

Hence n things can be arranged into s groups in

$$\frac{\lfloor\!\underline{n+s-1}}{\lfloor\!\underline{s-1}} \text{ ways.}$$

Example 320. In how many ways can 4 horses of the same sort be arranged in 3 stables?

Four horses can be arranged in 3 stables in

$$\frac{\underline{|4+3-1}}{\underline{|3-1}} \text{ ways.}$$

If the 4 horses are of the same sort they will have only $\frac{1}{\underline{|4}}$ as many arrangements, by Axiom 9.

Hence 4 horses of the same sort can be arranged in 3 stables in $\frac{\underline{|4+3-1}}{\underline{|4}\;\underline{|3-1}} = 15$ ways.

Similarly n like things can be arranged in s groups in

$$\frac{\underline{|n+s-1}}{\underline{|n}\;\underline{|s-1}} \text{ ways.}$$

Example 321. In how many ways can 4 like horses be put into 3 stables ?

Four like horses can be arranged in 3 stables in

$$\frac{\underline{|n+s-1}}{\underline{|n}\;\underline{|s-1}} \text{ ways, by Example 320.}$$

And since the horses are all alike they can have only one arrangement, by Definition 6.

Hence 4 like horses can be put into 3 stables in the same number of ways in which they can be arranged in 3 stables.

Hence 4 like horses can be put into 3 stables in

$$\frac{\underline{|4+3-1}}{\underline{|4}\;\underline{|3-1}} = \frac{\underline{|n+s-1}}{\underline{|n}\;\underline{|s-1}} = \frac{\underline{|6}}{\underline{|4}\;\underline{|2}} = 15 \text{ ways.}$$

Hence n like things can be put into s parcels in $\frac{\underline{|n+s-1}}{\underline{|n}\;\underline{|s-1}}$ ways, by Definition 4.

Example 322. If 10 men select a president by ballot from 3 men, in how many ways can the vote result?

$$\text{In } \frac{\underline{|10 + 3 - 1}}{\underline{|10}\,\underline{|3 - 1}} = \frac{\underline{|12}}{\underline{|10}\,\underline{|2}} = 66 \text{ ways.}$$

Example 323. If 9 ball players select a captain from among themselves by ballot, in how many ways can the vote result?

$$\text{In } \frac{\underline{|9 + 9 - 1}}{\underline{|9}\,\underline{|9 - 1}} = \frac{\underline{|17}}{\underline{|9}\,\underline{|8}} \text{ ways.}$$

Example 324. If 9 men select a captain from one man by ballot, in how many ways can the vote result?

Example 325. If 5 like men have 3 shacks, in how many ways can they occupy some of them?

Some of the shacks includes all of them.

$$\text{In } \frac{\underline{|5 + 3 - 1}}{\underline{|5}\,\underline{|3 - 1}} = \frac{\underline{|7}}{\underline{|5}\,\underline{|2}} = 21 \text{ ways.}$$

Example 326. In how many ways can 4 like horses be put into 3 stables if none of the stables are to be left empty?

Three like horses can be taken from 4 like horses in 1 way, by Definition 6.

And they can be put 1 into each of the 3 stables in 1 way, by Definition 6.

And the remaining 4 — 3 like horses can be put into the three stables in $\frac{\underline{|4 - 3 + 3 - 1}}{\underline{|4 - 3}\,\underline{|3 - 1}} = \frac{\underline{|4 - 1}}{\underline{|4 - 3}\,\underline{|3 - 1}} =$ 3 ways.

Hence the 4 like horses can be put into 3 stables so as to leave no blanks in $1 \times 1 \times \dfrac{(4-1)!}{(4-3)!\,(3-1)!} = 3$ ways, by Axiom 8.

Similarly n like things can be put into s parcels so as to leave no blanks in $\dfrac{(n-s+s-1)!}{(n-s)!\,(s-1)!} = \dfrac{(n-1)!}{(n-s)!\,(s-1)!}$ ways.

Example 327. If 5 like men have 3 shacks, in how many ways can they occupy them if no shack is to be left alone?

$$\text{In } \frac{(5-1)!}{(5-3)!\,(3-1)!} = \frac{4!}{2!\,2!} = 6 \text{ ways.}$$

Example 328. If 5 men live by chance in 3 shacks, what is the chance that a given shack will be empty?

Each of the men may live in any of the 3 shacks.

Hence the 5 men can live in the 3 shacks in 3^5 ways.

And they can live in any given 2 shacks in 2^5 ways.

Hence the chance that the other shack will be empty is

$$\text{is } \frac{2^5}{3^5} = \left(\frac{2}{3}\right)^5 = \frac{32}{243} = \cdot 1317.$$

Example 329. In how many ways can 4 like dimes be carried in 4 pockets?

Example 330. In how many ways can 4 like dimes be carried in 3 pockets with no blanks?

Example 331. In how many ways can 4 rings be worn on 3 fingers? See Example 319.

Example 332. In how many ways can 4 rings be worn on 3 fingers with no blanks ?

Four like rings can be put into 3 parcels with no blanks in $\frac{(4-1)!}{(4-3)!\,(3-1)!}$ = 3 ways, by Example 326.

But the 4 rings are different, by Definition 6.

Hence they can be arranged in $4!$ ways, by Axiom 9.

Hence 4 rings can be worn on 3 fingers so as to leave no blanks in $3 \times 24 = 72$ ways.

Similarly n things can be arranged in s groups with no blanks in $\frac{n!\,(n-1)!}{(n-s)!\,(s-1)!}$ ways.

Example 333. In how many ways can 4 men be arranged on 3 benches in a row ?—Ans. 72.

Example 334. In how many ways can 4 flags be arranged on 3 poles ?—Ans. 360.

Example 335 In how many ways can 4 flags be arranged on 3 poles with no blank poles ?—Ans. 72.

Example 336. In how many ways can 4 like flags be arranged on 3 poles ?—Ans. 15.

Example 337. In how many ways can 4 like flags be arranged on 3 poles with no blanks ?—Ans. 3.

Example 338. In how many ways can 4 like flags be put on 3 poles with no blanks ?—Ans. 3.

Example 339. In how many ways can 4 flags be put on 3 poles with no blanks ?—Ans. 36.

Example 340. In how many ways can 4 flags be arranged ?—Ans. 24.

Example 341. If there are 4 flags left on a pole and the order of the flags has been changed, what is the chance that the first flag still remains first ?

Four flags can be arranged in $\underline{|4}$ ways, and 3 flags in $\underline{|3}$ ways.

Hence the chance that the first flag was not changed is

$$\frac{\underline{|3}}{\underline{|4}} = \frac{1}{4}.$$

Example 342. What is the chance that the first 2 flags were not changed ?

The remaining 2 flags can be arranged in $\underline{|2}$ ways.

Hence the chance that the first 2 were not changed is

$$\frac{\underline{|2}}{\underline{|4}} = \frac{1}{12}.$$

Example 343. What is the chance that no flag remains in its original place ?

$(1 - p)^4 = 1 - 4p_1 + 6p_2 - 4p_3 + p_4$, by Proposition 2.

$$= 1 - 4\frac{\underline{|3}}{\underline{|4}} + 6\frac{\underline{|2}}{\underline{|4}} - 4\frac{\underline{|1}}{\underline{|4}} + \frac{\underline{|0}}{\underline{|4}}$$

$$= 1 - 1 + \tfrac{1}{2} - \frac{1}{6} + \frac{1}{24} = \frac{3}{8}$$

Hence the chance that no flag remains in the same place is $\frac{3}{8}$.

Example 344. If 4 men sit on 4 chairs in a row and later they sit by chance on the 4 chairs, what is the chance that none of them will sit on the same chairs as formerly ?—Ans. $\frac{3}{8}$.

Example 345. In how many ways can 4 flags on a pole be deranged so that no flag occupies its original place ?

Four Flags can be arranged in $\underline{|4} = 24$ ways.

The chance that no flag will occupy its original place is $\frac{3}{8}$.

Hence the 4 flags can be deranged so that no flag will occupy its original place in $\frac{3}{8}$ of $\underline{|4} = \frac{3}{8}$ of $24 = 9$ ways, by Definition 3.

Example 346. And in general if n things are deranged the chance that none of s given things will remain in their original places is

$$(1-p)^s = 1 - sp_1 + \frac{\underline{|s}}{\underline{|s-2}\,\underline{|2}}\,p_2 - \frac{\underline{|s}}{\underline{|s-3}\,\underline{|3}}\,p_3 + \text{etc.}$$

$$\text{where } p_1 = \frac{\underline{|n-1}}{\underline{|n}}$$

$$p_2 = \frac{\underline{|n-2}}{\underline{|n}}$$

$$p_r = \frac{\underline{|n-r}}{\underline{|n}}$$

And the number of ways in which n things can be deranged so that none of s given things will remain in their original places is

$$\underline{|n}\,(1-p)^s = \underline{|n} - s\,\underline{|n-1} + \frac{\underline{|s}}{\underline{|s-2}\,\underline{|2}}\,\underline{|n-2} - \frac{\underline{|s}}{\underline{|s-3}\,\underline{|3}}\,\underline{|n-3} + \text{etc.}$$

$$= \Sigma\,\frac{\underline{|s}}{\underline{|s-r}\,\underline{|r}}\,\underline{|n-r}\,(-1)^r.$$

When $s = n$ this is

$$\Sigma \frac{\underline{|n}}{\underline{|n-r}\,\underline{|r}} \underline{|n-r}\,(-1)^r = \underline{|n}\,\Sigma \frac{(-1)^r}{\underline{|r}}$$

$$= \underline{|n}\left(1 - 1 + \frac{1}{\underline{|2}} - \frac{1}{\underline{|3}} + \frac{1}{\underline{|4}} \cdots\cdots \pm \frac{1}{\underline{|n}}\right)$$

$$= \frac{\underline{|n}}{e}\text{, if } n \text{ is very large, by Proposition 2 and Def. 14.}$$

Example 347. If there are 4 flowers in a row and they have been deranged, what is the chance that neither of 2 given flowers occupy their original places?

It is $(1 - p)^2 = 1 - 2p_1 + p_2$

$$= 1 - 2\frac{\underline{|n-1}}{\underline{|n}} + \frac{\underline{|n-2}}{\underline{|n}} = 1 - \tfrac{1}{2} + \tfrac{1}{12} = \frac{7}{12}.$$

Example 348. In how many ways can the 4 flowers be deranged so that neither of the first 2 flowers will occupy its original place?

It is $\frac{7}{12}$ of $\underline{|4} = 14$.

Example 349. In how many ways can the 4 letters *abcd* be deranged so that neither *a* nor *b* will occupy its original place?—Ans. 14.

Example 350. What is the chance that *ab* will remain the same?

It is $p_2 = \dfrac{\underline{|n-2}}{\underline{|n}} = \dfrac{1}{12}$.

Example 351. What is the chance that 2 letters will remain the same?

It is $\dfrac{\underline{|4}}{\underline{|2}\,\underline{|2}} \cdot \dfrac{\underline{|n-2}}{\underline{|n}} = \tfrac{1}{2}$.

Example 352. In how many ways can 2 letters remain the same ?

In $\frac{1}{2}$ of $4! = 12$ ways.

Example 353. If there are r things of one sort and s things of another sort and t things of another sort in how many ways can a different combination be made from them all ?

Since all the r things are of the same sort a different combination can be made only by taking a different number of things, by Definition 4 and 6.

We can take 0, 1, 2, 3,. . . . . . r things for any combination. Hence a different combination can be made from the r things in $r + 1$ ways.

And from all the things in $(r + 1)(s + 1)(t + 1)$ ways, by Axiom 8.

And one of these is blank.

Example 354. The total number of arrangements of n things 0 at a time, 1 at a time, 2 at a time, 3 at a time, and so on, is $\Sigma \frac{n!}{(n-r)!}$, where r successively takes the values, $n, n - 1, n - 2, \ldots\ldots 0$.

$$\Sigma \frac{n!}{(n-r)!} = n!\left(1 + 1 + \frac{1}{2!} + \frac{1}{3} \ldots\ldots + \frac{1}{n}\right)$$

$= e\, n!$, if n is very large, by Proposition 2 and Definition 14.

And e is the total number of chances of a combination from anything when all the elements are very small, by Proposition 2, and Definition 2.

Hence the total number of arrangements of n things is equal to the number of arrangements of n things all at a time multiplied by the total number of chances of a combination from one thing composed of very small elements, by Definition 14.

Whitworth's *Choice and Chance* works out this part of the subject much more fully than we can here. All our proofs in this chapter are based on Proposition 2. There they are based on the possible arrangements of the things in the parcels and the partitions between the parcels.

CHAPTER XI

MISCELLANEOUS

Example 355. In times of epidemics, sinking of ships in time of war, accidents of all kinds that occur so many a week, events of interest to us that occur about so many times on an average in a certain length of time, it is very convenient to carry formulæ in the head that will enable us to ascertain approximately the chance that any number will occur, so as to give us an idea whether things are getting better or worse. In the world war when our ships were being sunk by submarines I was anxious to have such a formula, and tried to make one, and that of Proposition 7 is the result. I was surprised to find it so precise. It was so precise that it must represent the law of errors. But it was difficult to get the proof.

A simple approximate formula for the chance that an error will be less than sr when s is less than 1 is $P = \frac{s}{2}$. When s is 5 or greater, the chance that an error will be greater than sr is $1 - P = \frac{1}{5}\left(\frac{1}{2}\right)^{\frac{s^2}{3}}$, very nearly.

Example 356. If the Median error is 7 and the average *obs* of anything is 100, what is the chance of an *obs* of it of over 198 ?

The error is $x = 198 - 100 = +98 = +14r = +sr$, by Definition 14.

Hence the chance is

$$\frac{1 - P}{2} = \tfrac{1}{2}\left(\tfrac{1}{2}\right)^{\frac{s^2}{3}} = \cdot 000{,}000{,}000{,}000{,}000{,}000{,}01,$$ by Proposition 13.

Example 357. If the Median error in any series of *obs* is 12, what is the chance that the error will be greater than 36 ?

It is $\dfrac{\left(\frac{1}{2}\right)^2}{\underline{|3}} = \dfrac{1}{24} = \cdot 042$, very nearly, by Proposition 7.

The Tables in the back of the book give ·043.

Example 358. What is the chance that the error will be greater than 60 ?

$$\frac{(\frac{1}{2})^3}{\underline{|5}} = \frac{1}{8 \times 120} = \cdot 001,$$ by Proposition 7.

The Tables in the back of the book give ·00074.

$\frac{1}{5}(\frac{1}{2})^{\frac{s^2}{3}}$ gives ·00062, and is very close for values of s not less than 5.

Example 359. What is the chance that the error will be greater than 48 ?

$$\frac{\left(\frac{1}{2}\right)^{\frac{s+1}{2}}}{\underline{|s}} = \cdot 707 \frac{\left(\frac{1}{2}\right)^{\frac{s}{2}}}{\underline{|s}} = \cdot 0074.$$

The Tables give ·00698.

Example 360. What is the chance that the error will be greater than 30 ?

$30 = 12 \times 2{\cdot}5 = 2{\cdot}5r = sr.$

$|\underline{2\cdot5} = 2\cdot5 \times 1\cdot5 \times |\underline{\cdot5} = |\underline{n} = n\ (n - 1)\ |\underline{n - 2}.$

$|\underline{\cdot5} = \cdot89$, by Example 109.

Hence $|\underline{2\cdot5} = 2\cdot5 \times 1\cdot5 \times \cdot89 = 3\cdot32$.

Hence $1 - \mathrm{P} = \dfrac{\left(\frac{1}{2}\right)^{\frac{s+1}{2}}}{|\underline{s}} = \cdot09.$

Example 361. What is the value of $|\underline{\cdot3}$?

$$1 - \mathrm{P} = \frac{\left(\frac{1}{2}\right)^{\frac{s+1}{2}}}{|\underline{s}}.$$

This formula is very close foi values of s from 1 to 2.

Take $s = 1\cdot3$.

$1 - \mathrm{P} = \dfrac{\left(\frac{1}{2}\right)^{1.15}}{|\underline{1.3}} = \cdot3806$, by the Tables.

Hence $|\underline{1\cdot3} = 1\cdot3 \times |\underline{\cdot3} = 1\cdot184 = |\underline{n} = n\ |\underline{n - 1}.$

Hence $|\underline{\cdot3} = \cdot91$, by Definition 14.

Example 362. The Principal of an Edmonton, Alberta, school, Mr. W. H. Todd, measured the quality of intellects of a class of 38 by the Binet-Simon scale and found the average quality of the class to be 108, and the average variation to be 14. What is the normal error?

$x_0 = x_a\sqrt{\dfrac{\pi}{2}} = 1\cdot2533x_a = \dfrac{5}{4}\,x_a = \dfrac{5}{4} \times 14 = 17\cdot5,$

by Definition 14. See Example 118.

The Normal error was 17.

Example 363. What is the Median error?

It is $\cdot8454x_a = \dfrac{6}{7}\,x_a = \dfrac{6}{7}$ of $14 = 12$, by Definition 14.

The Median error was between 12 and 13.

Example 364. Taking this as a standard what is the chance that an intellect will be over 168 ?

The residual error is $168 - 108 = +60 = 5r$

Hence the chance is $\frac{1}{2}(1 - \mathrm{P}) = \frac{1}{2}\,\frac{(\frac{1}{2})^3}{\underline{|5}} = \frac{1}{1920} =$ ·0005, by Definition 14.

Example 365. What is the chance of an intellect under 72 ?

The error is $72 - 108 = -36 = 3r$, by Definition 14.

$$\frac{1 - \mathrm{P}}{2} = \frac{1}{2}\,\frac{\left(\frac{1}{2}\right)^{\frac{s+1}{2}}}{\underline{|s}}$$

$$= \frac{\left(\frac{1}{2}\right)^{3}}{\underline{|3}} = \frac{1}{48} = \cdot 02.$$

Example 366. What is the chance of an intellect below 48 ?

The error is $48 - 108 = -60 = 5r$, by Definition 14.

Hence the chance is ·0005.

Example 367. What is the chance of an intellect over 142 ?

The error is $142 - 108 = +34$, by Definition 14.

$$t = \frac{x}{x_0\sqrt{2}} = \frac{x}{17{\cdot}5 \times 1{\cdot}414} = \frac{4x}{100} = 1{\cdot}36.$$

$$\frac{1 - \mathrm{P}}{2} = \cdot 027 = \frac{1}{37}\,.$$

Hence there should be one over 142 in the group.

There were 2 each 142, and none above 142.

Example 368. How many in the class should be below 74 ?—Ans. 1.

There was 1 below 74.

Example 369. How many in the class should be above 128 ?

$$t = \cdot 8.$$

$$\frac{1 - P}{2} = \cdot 13.$$

$$\cdot 13 \times 38 = 5.$$

There were 5 above 128 in the class.

Example 370. How many in the class should be above 110 ?

$$t = \frac{4x}{100} = \cdot 08.$$

$$\frac{1 - P}{2} = \cdot 455.$$

$$\cdot 455 \times 38 = 17.$$

There were 17 above 110 in the class.

Example 371. How many in the class should be below 90 ?

$$t = \cdot 72.$$

$$\frac{1 - P}{2} = \cdot 155.$$

$$\cdot 155 \times 38 = 5 \cdot 9.$$

There were 5 below 90 in the class.

Example 372. Since the Median error is 12 one half of the class should be from 108 + 12 = 120 to 108 — 12 = 96.

There were 19 from 96 to 120.
$\frac{1}{4}$ or $9\frac{1}{2}$ should be below 96.
There were 11 below 96.
$\frac{1}{4}$ or $9\frac{1}{2}$ should be from 108 to 120
There were 10 from 108 to 120.

$\frac{1}{4}$ or $9\frac{1}{2}$ should be above 120.

There were 8 above 120.

In the *Measurement of Intelligence by the Binet-Simon Scale* Professor Terman gives the average for the whole country as about 100, and the Median error as about 7.

This class of children observed by Mr. Todd were in a higher grade to which many dull children do not attain, which accounts for the high average and Median error.

If the average quality of the whole country is 100 and the Median error 7 the chance of an average of 108 or more for 38 persons is the chance of an error of more than $\sqrt{8^2 \times 38} = 8\sqrt{38} = +49$ for a single person, by Definition 13.

$49 = 7r$.

Hence the chance is ·000,001.

This very small chance indicates that many of the children were bright.

Example 373. What is the average error for the whole country?

Example 374. What is the Normal error?

Example 375. What is the chance of an intellect whose quality is over 168?

The error is $168 - 100 = +68$, by Definition 14.

$$t = \frac{x}{x_0\sqrt{2}} = \cdot0676x = 4\cdot60.$$

$$\frac{1 - P}{2} = \cdot000,000,000,04.$$

Example 376. If in a normal examination of a class of 48 students half are in Class C, $\frac{3}{16}$ in Class B, $\frac{3}{16}$ in Class D, $\frac{1}{16}$ in Class A and $\frac{1}{16}$ in Class E, who fail, what is the chance that one or more will fail through bad luck?

Normally there are 3 in Class E.

The chance that one will fail is $\frac{1}{16}$ and the chance that he will not fail is $\frac{15}{16}$.

Hence the normal number of chances of a pair of contradictories is $2\pi npq = 2\pi \times 3 \times \frac{1}{16} \times \frac{15}{16}$.

The normal error is $\sqrt{3 \times \frac{1}{16} \times \frac{15}{16}} = .42$.

$$t = \frac{x}{x_0\sqrt{2}} = 1.7.$$

$\frac{1 - P}{2} = .01$, which is the answer, by Def. 14.

The chance that one or more will pass through good luck is the same.

The chance that one or more will get into Class A through good luck is the same, and the chance that one or more will not get into Class A through bad luck is the same.

Example 377. What is the chance that one or more who belong to Class C will get into Class B ?

Example 378. What is the chance that one or more who belong to Class B will get into Class C ?

Example 379. What is the chance that one or more who belong to Class A will be plucked ?

Example 380. What is the chance that one or more who belong to Class D will be plucked ?

Example 381. What is the chance that there will be over 15 at church if the expectation is 10 and the Median error 2 ?

Example 382. What is the chance that there will be over 1500 at church if the expectation is 1000 and the Median error 2 ?

N

Example 383. What is the chance that a horse will win by over 25 feet if his expectation is 10 feet and his average error 50 feet ?

Example 384. Show that the normal error in s dimensions of space is equal to $x_0\sqrt{s}$, and that the most likely error is equal to $x_0\sqrt{s-1}$, where x_0, is the normal error in any dimension. See Examples 186 and 532.

The Median error is expressed approximately by $x_0\sqrt{as+b}$. What are the most likely values of a and b?

Example 385. If A and 5 others each stake 25 cents in a pool on a race, what is the chance that he will win more than $2 in 10 races ?

Example 386. If a line is observed 10 times and the average error is 2, what is the chance that in observing it again the error will be over 6 inches ?

Example 387. What is the average error of the average length ?

Example 388. If anything is observed 5 times under one set of conditions and its Median value is 1103 ± ·31, and it is also *obs* 5 times under another set of conditions, and its Median value is 1102 ± ·123, what is its Median value ?

Example 389. If a line is *obs* twice under the same conditions and the Median length is 347 ± ·06, what is the Median length of a single *obs* ?

Example 390. What is the Median error of a line of 100 feet under the same conditions ?

Example 391. If anything is *obs* under one set of conditions and its Median value is 1013 ± 1 and it is also *obs* under another set of conditions and its Median value is 1012 ± ·02, how many times must it be *obs* under the former set in order to give as good a result as 2 *obs* under the latter ?

Example 392. If the 3 angles of a triangle are *obs* twice under the same set of conditions and the values got are

A	B	C
27° 41′	96° 24′	55° 54′
27° 43′	96° 25′	55° 51′

what is the average error of A, B and C, respectively ? See Definition 13.

Example 393. What is the average error of a single angle ?

Example 394. What is the Median value of A, B and C, respectively ?

Example 395. What are the most likely values of A, B and C, respectively ?

Example 396. What is the weight of a single *obs* of A, B and C, respectively ? See Definition 13.

Example 397. How many times would an angle have to be *obs* under a set of conditions whose average error for a single *obs* is ·12′ in order that the result would be as good as 100 *obs* made under the former set of conditions ?

Example 398. If A's chance to win any game is ·2 and there are 6 players each staking 10 cents a game, what is the chance that A will win more than $2 in 100 games ?

Example 399. What is the chance that he will lose more than $2 in 100 games ?

Example 400. In how many ways can a bouquet of 4 flowers be made from 6 kinds of flowers ?

Example 401. In how many ways can a bouquet of 4 flowers be made from 6 flowers ?

Example 402. If a set of dominoes go from double 0 to double 10, how many dominoes are there in the set ?

Example 403. If there are 4 horses in a stable and the chance that any one will be taken is $\frac{1}{2}$, that any 2 will be taken is $\frac{1}{2}$, that any 3 will be taken is $\frac{1}{2}$ and that all will be taken is $\frac{1}{2}$, what is the chance that none will be taken ?

Example 404. If the chance that any one will be taken is $\frac{3}{4}$, that any 2 will be taken is $\frac{1}{2}$, that any 3 will be taken is $\frac{1}{4}$, and that all will be taken is $\frac{1}{8}$, what is the chance that none will be taken ?

Example 405. If the chance that any one is true is $\frac{3}{4}$, that any 2 are true is $\frac{1}{2}$, that any 3 are true is $\frac{1}{4}$, and that all are true is $\frac{1}{8}$, what is the chance that none of the 4 statements are true ?

Example 406. If the chance that any one is false is $\frac{3}{4}$, that any 2 are false is $\frac{1}{2}$, that any 3 are false is $\frac{1}{4}$, and that all are false is $\frac{1}{8}$, what is the chance that all the statements are true ?

Example 407. If Leap year came by chance and the expectation was that it would come once in 4 years, what is the chance that it would come once in the next 4 years?

It is the chance of no error in one period.

When n is very large this is $\frac{1}{\sqrt{2\pi}} = \cdot 4$, by Example 29 and Definition 14.

Example 408. What is the chance that it would come every year for the next 4 years ?

It is $\frac{1^4}{\lfloor 4} e^{-1} = \cdot 015$, by Definition 14.

Example 409. What is the chance that it would come 3 times in the next 7 years ?

In 7 years there are $\frac{7}{4}$ periods of 4 years each.

Hence it is $\frac{\left(\frac{7}{4}\right)^3}{\lfloor 3} e^{-\frac{7}{4}} = \cdot 155$, by Definition 14.

Example 410. If Sunday came by chance and the expectation was that it would come once a week what is the chance that to-morrow would be Sunday ?

Example 411. What is the chance that next Sunday would be Sunday ?

Example 412. If in a store the expectation is to take in $100 a day and the average error is 12, what is the chance that over $3000 will be taken in in the next 25 days ?

Example 413. What is the Median error ?

Example 414. What is the average sales a day ?

Example 415. What is the normal error a day ? A week ? A month ?

Example 416. In how many ways can 4 apples and 5 oranges be given to 2 boys ?

Example 417. So that each will get something ?

Example 418. So that each will get at least one orange ?

Example 419. So that each will get at least one apple and one orange ?

Example 420. If the average error of a man is 10, what is the chance that he will make an error of 20 ?

Example 421. If the average error of a man is 2, what is the chance that he will make an error of 20 ?

Example 422. If a judge's average error is 5 he will often make an error of 10, but very seldom one of 20, by Axiom 3.

If his average error is 2, what is the chance that he will make an error of 10 ? Of 20 ?

He will never make an error of 20.

Example 423. If a man has $500 and he plays Roulette at Monte Carlo by staking $1 each game on single numbers, what is the chance that he will be ruined some time during 1000 games ? What is his most likely error some time during the series ?

Example 424. If a man whose median error at Target practice at a range of 100 yards is 4 inches, shoots 100 shots at a bull's-eye 1 ft. 4 ins. in diameter at a range of 200 yards, what is the chance that he will make 50 hits ?

Example 425. What is the chance that he will make 60 hits ? 80 hits ? 100 hits ?

Example 426. If a man shoots 100 shots at a range of 600 yards and puts 50 inside a ring of 4 ft. diameter, what is his median error at a range of 100 yards ?

Example 427. If a man shoots 17 shots at a range of 100 yards, and the distance from the centre to his 9th best shot is 4 inches, what is his median error at a range of 600 yards ?

Example **428.** If a line is *obs* 4 times and the lengths got are **1043·6**, **1043·2**, **1043·7** and **1043·3**, what is its most likely length ?

What is the average error ? What is the Median error ? What is the chance of an error of more than ·5 in a single *obs* ? What is the chance of an error of the average length of more than ·2 ? What is the normal error as found from the average error ? As found from the sum of the squares ?

Example **429.** In how many ways can 6 horses be arranged in 3 stables ?

Example **430.** In how many ways can 6 horses be put into 3 stables ?

Example **431.** In how many ways can 6 horses be put into 3 stables with no blanks ?

Example **432.** In how many ways can 6 horses be put into 3 stables with one in each of 2 stables ?

Example **433.** If 6 horses are put into 3 stables by chance, what is the chance that there will be no blanks ?

Example **434.** If in an election for an office there are 4 candidates and 10 voters, in how many ways can the vote result if the votes are all of the same sort as in vote by ballot ?—Ans. 286.

Example **435.** If by open vote as in the old days ?—Ans. $286\ \underline{|10}$.

Example **436.** What is the ratio of the median error in *obs* and that in Target practice if the Normal error in any dimension is 2 ?

Example **437.** If anywhere there are 2 contradictories, yes and no, how many dimensions of space must there be there ?—Ans. 3.

Example 438. If in Fairyland there are 3 contradictories, how many dimensions of space must there be there ?—Ans. 7.

Example 439. If somewhere there are *s* contradictories, how many dimensions of space must there be there ?—Ans. $2^s - 1$.

Example 440. In how many ways can 4 cents, 4 nickels, and 4 dimes be given to 3 boys if the coins are all marked as different ?

Example 441. In how many ways can they be given so that each boy will have some ?

Example 442. In how many ways if A is to have only cents, B only nickels and C only dimes ?

Example 443. If the cents, nickels and dimes are all alike, respectively ?

Example 444. If a line is *obs* 9 times and its average length is 1000 and the Median error of a single *obs* is ·6 what is the Median length of the line ?

Example 445. What is the average error ? The normal error ?

Example 446. What is the chance that in 9 more *obs* the average length will be more than 1000·4 ?

Example 447. If a man puts half his shots in a circle of 1 inch diameter, what is his median error at that range ?

Example 448. What is the chance that he will put half his shots in a circle of one inch radius at twice the range ?

Example 449. If in shooting at an airplane the median error at a mile is 40 feet, what is the chance that a shell will burst within 40 feet of its centre at a range of 2 miles ?

Example 450. If when 2 men are playing Pool the average cash error is $4, what is the average cash error when 6 men are playing ?

Example 451. In playing Roulette at Monte Carlo, what difference does it make whether a player takes back half his stake or leaves it in prison when 0 turns up ?

Example 452. In how many ways can a pack of 52 cards be given to 4 players ?

Example 453. So that each will have a combination of 13 cards ?

Example 454. So that each will have a group of 13 cards ?

Example 455. So that A will have none ?

Example 456. So that neither A nor B will have any ?

Example 457. So that each will have some ?

Example 458. So that A will have all ?

Example 459. In how many ways can 52 cards be arranged if they are all different ?

Example 460. If there are 13 of each of 4 sorts ?

Example 461. In how many ways can 52 cards be divided up into 4 combinations of 13 cards each ?

Example 462. Into 4 combinations of 10, 15, 20 and 7, respectively ?

Example 463. If the average daily attendance at a school is 100 and the average variation is 2, what is the chance of less than 80 any day ?

Example 464. How many trials at Head or Tail must there be in a series in order that the chance of an error of 25 some time during the series will be 4 times as great as the chance of an error of 25 at the end of the series ?

Example 465. What is the chance that a hand of 5 cards will be 3, 4, 5, 6 and 7 from a pack of 52 cards ?

Example 466. If the Median height of an Englishman is 67·375 ± 1·73 inches, what is the chance of an Englishman over 84.675 inches in height ?

Example 467. If in any district the chance of blindness is 10 times the chance for the whole country, we would feel sure that the conditions there were not normal, and the government would send men there to find out what was wrong, or if two or more evils each occur alone but seldom, but together often. These investigations often lead to important knowledge as to the nature of things.

Example 468. If a line is *obs* 9 times and its Median length is 26347 ± ·03, what is the chance that in a single *obs* its length will be over 26349 ?

Example 469. In how many ways can 4 men be divided up into 4 combinations of 1 man each ? Into 2 combinations of 2 men each ?

Example 470. If there are 4 men of whom 2 are of one sort and 2 of another sort, in how many ways can they be arranged ?

Example 471. If there are 4 men and their wives, in how many ways can 4 couples be formed ?—Ans. 24.

Example 472. In how many groups of 8 can 4 men and their wives enter a room ?

Example 473. If the men are to be together and the women together?

Example 474. If no 2 men or 2 women are to be together.—Ans. 2880.

Example 475. If they are to enter in couples?—Ans. 576.

Example 476. If each man must enter with his wife? —Ans. 24.

Example 477. If no man is to enter with his wife?

Example 478. If they are to enter 4 at a time?

Example 479. If each 4 must be 2 men and 2 women?

Example 480. If each 4 must be all men or all women?

Example 481. If each 4 must be 3 men and 1 woman or 3 women and 1 man?

Example 482. In how many ways can a combination of 4 be made from the 8 persons?

Example 483. In how many ways can the 8 persons be divided up into 4 combinations of 2 persons each?

Example 484. Into 4 combinations of 1, 2, 3 and 2, respectively?

Example 485. Into 4 combinations each composed of a man and a woman?

Example 486. In how many ways can a combination composed of 3 women and 1 man be made from the 8 persons?

Example 487. In how many ways can a combination of 2 women and 1 man be made from the 8 persons?

Example 488. If a soldier's median error at a range of 100 yards is 3 inches, what is the chance that he will put 50 shots out of 100 in a ring of 6 inches diameter at that range?

The chance that he will put any shot in the ring is $p = \frac{1}{2}$, and the chance that he will not is $q = \frac{1}{2}$.

Hence the chance that he will put 50 out of 100 is

$$\frac{100!}{50!\,50!} p^{50} q^{50} = \frac{100!}{50!\,50!} (\tfrac{1}{2})^{100} = \frac{1}{\sqrt{2\pi n p q}} = \cdot 08.$$

Example 489. What is the chance that he will put 100 in the ring? 55 in the ring?—Ans. $(\frac{1}{2})^{100}$, ·05.

Example 490. If there are 8 things of one sort and 4 things of another sort in how many ways can a different combination be made?—Ans. 45, and one of these is blank.

Example 491. In how many ways can 4 peaches and 2 apples be given to 3 boys?

Example 492. In how many ways can a combination be made from 4 peaches and 2 apples?

Example 493. In how many ways can 4 peaches and 2 plums be arranged?

Example 494. If 4 peaches and 2 plums in a row measure 9 inches, and 6 peaches and 4 plums 15 inches, and 3 peaches and 2 plums 7 inches, and 8 peaches and 9 plums 24 inches, and 15 peaches and 3 plums 32 inches, and 40 peaches and 60 plums 12 feet, what is the most likely diameter of a peach? Of a plum?

What is the chance that a peach will be over 3 inches in diameter? What is the chance that a plum will be over 2 inches in diameter?

Example 495. If one series of *obs* gives the elevation of A as 3002 feet above the sea, and another series of *obs* gives it as 3003, what is the most likely elevation of A if the weights of the first and second series are 1 and 25, respectively ?

Example 496. If one series of *obs* gives 3A + 4B = 21,000, and another series gives 2A + 10B = 36,001, and another A — B = ·5, what are the most likely elevations of A and B, respectively, if the weights of the series are 1, 4, and 100 ? What is the weight of A ? What is the chance that in another *obs* A — B will be over 1 ?

Example 497. The definition of a dimension of space is, a dimension of space is that in which, if a force is acting in the direction of the dimension, the force has no component in the direction of any other dimension of space.

If anywhere there are 4 dimensions of space, how many contradictories must there be there ?—Ans. 2·3219.

Example 498. If the weight of a single *obs* is 1 and the *obs* equations are $2^x = 7$, $3^x = 25$ and $4^x = 60$, what is the most likely value of x ? What is its weight ? What is its weight in the first equation ? What is the chance that the value of x is greater than 3.

Example 499. If 1 out of 30 children is a twin what is the chance that there will be 3 or more pairs of twins among any 100 children ?

Example 500. In Example 204 what was the most likely error some time during the *obs* ?

Example 501. There is a record that A drew a white ball first draw from a bag containing 1,000 balls of which one was white. The chance that he was not mistaken in the number and nature of the balls is p_1 = ·99, the chance that he read his ball correctly is $p_2 = \cdot 99$, the chance that his record at that time was faithful is $p_3 = \cdot 99$ and the chance that the record has not been changed is $p_4 = \cdot 99$. What is the chance that the record is true ?—Ans. $p_1p_2p_3p_4 + p_1p_2q_3q_4 + p_1q_2q_3p_4 + p_1q_2p_3q_4$.

Example 502. In a book the expectation is 10 words to a line and the average variation in any line is 1·5 words, what is the chance that any line will have more than 15 words ?

How many lines must there be in the book in order that we should expect 3 lines each containing more than 15 words ?

Example 503. If there are two statements and the chance that any one of them is true is $\frac{3}{4}$, show that the chance that they are both true cannot be less than $\frac{1}{2}$.

Example 504. If there are two statements, both of which cannot be true, and the chance that any one is true is $\frac{1}{2}$, show that they cannot both be false.

Example 505. In the north the sun distinctly whitens the south side of 9 out of 10 smooth barked Poplar trees. What is the chance that 50 or more out of any 100 will indicate north ? The north side is never white.

Example 506. The north side of 99 out of 100 Jack Pine trees have very small long tufts of moss scattered over its surface and extending from 1 to 5 feet from the ground. This can be distinctly seen from a distance of 100 feet. What is the chance that in a clump of 10 Jack Pine trees 5 or more will indicate north ?

Example 507. What is the ratio between Σx and Σv ? Between Σx^2 and Σv^2 ? When is this ratio equal to 1 ? When is it very nearly equal to 1 ? When is the average value equal to the expectation ?

Example 508. When n is large the Median error for one trial is $r = \frac{11}{13}$ of $\frac{\Sigma v}{n}$, nearly. What is the Median total error for a series of n trials?

Example 509. If $\frac{\Sigma v}{n} = 3{\cdot}9$, what is the chance of a residual error greater than 33? Than 13·2? Than 1·4? When is there no error in the average value?

Example 510. If in a district of 1,000,000 population there are 100,000 automobiles, what is the most likely number of automobiles in a similar district of 100,000,000 population? What is the normal variation? The Median variation? If there is a tax of $1 on each automobile, what is the chance that the tax will give less than $9,950,000 in the latter district?

Example 511. What is the value of $\frac{\Sigma x}{\Sigma v}$? Of $\frac{\Sigma x^2}{\Sigma v^2}$?

Of $\frac{x_0}{x_a}$? Of $\frac{x_0}{r}$? Of $\frac{r}{\Sigma v}$? Of $\frac{x_0}{\Sigma v}$? Of $\frac{x_a}{\Sigma v}$? Of $\frac{pv^2}{x_0{}^2}$?

From what is the error measured? See Definition 11.

Example 512. If a person passing through Monte Carlo thought he would like the experience of risking $100 at Rouge-et-Noir, or red and black at Roulette, what would be the best way to do it?

The best $\frac{3}{4}$ of all the shots of rifle fire is regarded as the *effective fire* in war. Hence if x is the greatest effective error $e^{-\frac{x^2}{2x_0{}^2}} = \frac{1}{4}$.

Hence $x = \frac{5}{3}x_0$.

And any error greater than $\frac{5}{3} x_0$ is not considered effective, and is disregarded, in war.

Hence in analogy to the science of war he should consider all chances less than $\frac{1}{4}$ as not effective, and disregard them.

The chance to lose more than x some time during a series of n games is $\frac{1}{2}e^{-\frac{x^2}{2nx_0^2}}$, by Proposition 13.

Hence $\frac{1}{2}e^{-\frac{x^2}{2nx_0^2}} = \frac{1}{4}$.

Hence $e^{-\frac{x^2}{2nx_0^2}} = \frac{1}{2}$.

Hence his greatest effective error some time during the series is the median total error at target practice, and is $x = 1{\cdot}1774 x_0\sqrt{n}$, where $x_0\sqrt{n}$ is the normal total cash error at the end of a series of n games.

Hence, if he disregards all chances of losing his \$100 that are disregarded in the science of war,

$$1{\cdot}1774 x_0\sqrt{n} = \$100.$$

Hence $x_0\sqrt{n} = \$85$.

His adverse expectation is $\frac{na}{74}$.

He should not allow his adverse expectation to be more than he is willing to pay for his seat at the table.

To sit at the table of a prince is worth \$25.

Hence $\frac{na}{74} = \$25$.

We have shown that when he plays on red or black $x_0 = a$, or his stake.

Hence $a\sqrt{n} = \$85$.

Hence $a = \$5$ and $n = 289$.

Thus in analogy to the science of war he should stake \$5 each play and plan on a series of 289 plays.

His most likely gain some time during the series is $x_0\sqrt{n} = \$85$.

The chance that he will win more than this some time during the series is ·3.

Hence the odds are 3 to 7 that he will win more than \$85 some time during the series.

If he is lucky enough to be \$85 ahead at any time he may stop.

But if he wants to take all the effective chances of war he will play on in the hope of winning \$100.

The chance that he will win more than \$100 some time during the series is $\frac{1}{4}$.

The odds are 1 to 3 that he will.

If he is lucky enough to be \$100 ahead at any time he should stop, for all larger errors are regarded as stray shots in war.

If he plans to play through the whole series of 289 plays, and to neglect the principles of war, he has a chance of $\frac{1}{4}$ to win more than the Median total error at the end of the series, or

$$r\sqrt{n} = \frac{2}{3}\,x_0\sqrt{n} = \$57.$$

If he has not been lucky and lost his \$100, or has made big gains, he may persist in playing more.

If he plays 1,000 games his adverse expectation is $\frac{na}{74} = \$68$.

For 1,000,000 games it is \$67,568.

This is what he pays for his seat.

A man would not expect to win if he bet on a horse whose odds to win were 1 to 99, but here fabulous fortunes rise before his eyes when the odds are infinitesimal.

The Median error at the end of a series of n games is $r\sqrt{n} = \frac{2}{3} x_0 \sqrt{n}$.

The chance to win more than this is $\frac{1}{4}$, giving odds of 1 to 3.

The chance to win more than $3r\sqrt{n}$ is $\frac{1}{2} \frac{\left(\frac{1}{2}\right)^2}{\lfloor 3} = \frac{1}{48}$, giving odds of 1 to 47.

The chance to win more than $7r\sqrt{n}$ is $\frac{1}{2} \frac{\left(\frac{1}{2}\right)^4}{\lfloor 7} = \frac{1}{816480}$, giving odds of 1 to 816479.

If $n = 10{,}000$ the Median total error is

$$r\sqrt{n} = \$333, \text{ and}$$
$$3r\sqrt{n} = \$999, \text{ and}$$
$$7r\sqrt{n} = \$2333.$$

The cost of the seat is \$676.

If $n = 1{,}000{,}000$ the Median total error is

$$r\sqrt{n} = \$3{,}333.$$
$$3r\sqrt{n} = \$9{,}999,$$
$$7r\sqrt{n} = \$23{,}333.$$

The cost of the seat is \$67,568.

The odds that he will win more than enough to pay for his seat are not 1 to 3, which are the least that are considered in war, but are infinitesimal.

The cost of his seat is about $21r\sqrt{n}$.

The chance to win more than this is $\frac{1}{2}(\frac{1}{2})^{\frac{(21)^2}{3}}$
$= \cdot 000,000,000,000,000,000,000,000,000,000,000,000,000,000,01$.

He has the same chance to win more than enough to pay for his seat as a blind man has to find his friend among 90,000,000,000,000,000,000,000,000,000,000,000,000,000,000 men by picking the first man he meets.

Example 513. In Example 204 the greatest number of men are about 67·5 inches in height.

Hence the expectation is about 67·5, by Proposition 12 and Definition 11. The formulæ of Natural chance applied to Common chance are more precise near the expectation, because the elements are smaller there.

Normally half the errors are less than the Median error. There are 6,194 measurements.

Hence normally $\frac{6194}{2} = 3097$ are less than the Median error. There are $918 + 881 + 886 = 2,685$ men from 66 to 69 inches in height.

Hence there are $3,097 - 2,685 = 412$ more men less than the Median height.

That will extend to about another ·25 inch.

Hence the Median error is about $\frac{69 - 66}{2} + \cdot 25 = 1 \cdot 75$.

This is nearly the same as we got in Example 204, and is close enough for many purposes.

The chance that a man will be over 5 feet 11 inches is the chance of an error of more than $+ 2r$, or $\frac{1}{2}$ of $\cdot 707 \frac{\frac{1}{2}}{\underline{|2}}$
$= \cdot 09$, by Definition 14.

Example 514. Show that the chance of an error greater than $x = sr$ sometime during any series, when the elements are all very small, is $(\frac{1}{2})^{\frac{s^2}{5}}$, approximately.

Any thing is a series by Definition 7.

Hence the chance is $e^{-\frac{x^2}{2x_0^2}}$, by Proposition 13.

$$e^{-\frac{s^2r^2}{2x_0^2}} = e^{-\frac{s^2}{4\cdot 4}} = (\tfrac{1}{2})^{\frac{s^2}{3}}, \text{ approximately.}$$

Example 515. What is the chance that some time during any series of 10,000 flips of a coin there will be 101 or more too many heads for that stage of the series ?

$x_0 = \sqrt{npq} = 50.$

$r = \cdot 6745x_0 = 33\cdot 7.$

The error is $+ 101 = sr = 3r$, by Definition 11 and Axiom 3.

Hence the chance is $\frac{1}{2}\,(\frac{1}{2})^{\frac{s^2}{3}} = \frac{1}{16}$, by Definition 14.

Example 516. If the Median cash error for any series is r, what is the chance of a gain of $9r$ or more some time during the series ? At the end of the series ?

Example 517. If the Median error for any measurement is r, what is the chance of an error of $15r$ or greater some time during any measurement ? At the end of any measurement ?

When $x = 11r$ or greater the formula for the chance some time during the series should be used for both, since when the error is very large it is almost sure to be very near the end of the series, and this formula has been proved for all values of *s*, and the chance of an error of *x* or more at the end of a series cannot be greater than its chance some time during the series, because the end of the series is some time during it. When the error is very large the two chances are practically the same. Compare with the Tables in the back of the book.

Example 518. If the Median error for any series is 2, what is the chance of an error of 100 or more some time during the series? At the end of the series?

Example 519. If in any city the average number of deaths a week from the plague is 100, and the average variation from this is 5, what is the chance of less than 10 deaths in any week? Of less than 15 in each of two weeks in succession? The average is for eight successive weeks.

If either of these events occurred, what would be the inference? See Example 118.

Example 520. If we do not know the general *obs* equation of a series of *obs* we can find it by inspection of the *obs* in simple cases, or by comparing the series of *obs* with Tables of logs, powers, sines, etc. And in more complicated cases we can plot the *obs* by rectangular co-ordinates and see what general curve passes near all the points.

If 1 man makes 21 toys in a day,

2 men make	43
3	62
4	79
10	198
20	404

the general *obs* equation is evidently $y = ax + b$, where $a = 20$, very nearly.

What are the most likely values of a and b? What is

the chance that 25 men would make more than 550 toys in any day? What is the chance that they would make less than 480? If the weight of a single *obs* is 1, what is the weight of a? Of b? Of the last of the 6 *obs*?

Example 521. If when $x = 1$, $y = 2$,
$x = 2$, $y = 4$,
$x = 6$, $y = 66$,

what is the general *obs* equation? What are the most likely values of the constants? What are the respective weights of the three *obs* if that of a single *obs* is 1?

Example 522. If when $x = 1$, $y = 0$,
$x = 4$, $y = \cdot 61$,
$x = 10$, $y = 1$,
$x = 100$, $y = 2$,

what is the general *obs* equation? What are the most likely values of the constants?

Example 523.

$$\begin{aligned} \text{If } ax^2 + by^2 &= 2\cdot 1 \text{ when } x = 1 \text{ and } y = 1, \\ &= 5\cdot 1 \quad x = 1 \quad y = 2, \\ &= 5\cdot 0 \quad x = 2 \quad y = 1, \\ &= 102\cdot 0 \quad x = 10 \quad y = 1, \end{aligned}$$

what are the most likely values of a and b? What is the chance that b will be greater than 1·3? If the weight of a single *obs* is 1, what are the respective weights of the 4 *obs*? What is the weight of a in the first equation? In the last? What is the weight of b in the last equation? What is the difference between the weight of a and its weight in the last equation? What is the difference between the weight of b and its weight in the last equation? See Example 186.

Example 524 If $a\,(\tfrac{1}{2})^x = \cdot 51$ when $x = 1$,
$= \cdot 26 \quad x = 2$,
$= \cdot 12 \quad x = 3$,

what is the most likely value of a ?—Ans. 1·021. The most likely general *obs* equation is $1{\cdot}021(\tfrac{1}{2})^x = k$.

What is the Median error for a ? What is the chance that a will be greater than 1·2 ? What is the weight of a in the first equation ? In the third ?

Example 525. If $a + b \log x = 1{\cdot}3$ when $x = 2$

$$\begin{aligned} &= 1{\cdot}9 \quad & x &= 8 \\ &= 3{\cdot}0 \quad & x &= 100, \end{aligned}$$

what are the most likely values of a and b ? What is the chance that a and b will each be greater than 1·2 ? What is the weight of b in the first equation ? In the third ? What is the difference between the weight of b and its weight in the last equation ?

Example 526.

$$\begin{aligned} \text{If } a + b \log x + cx^4 &= 2{\cdot}1 \text{ when } x = 1, \\ &= 17{\cdot}2 \qquad x = 2, \\ &= 83{\cdot}0 \qquad x = 3, \\ &= 258{\cdot}0 \qquad x = 4, \end{aligned}$$

what are the most likely values of a, b and c ? What is the chance that c will be less than 1·2 ? What is the weight of c in the first equation ? In the last ? The most likely value of c is controlled by the last equation.

Example 527. Reproduction and decay take place by the law of combinations.

The reproduction of plants is a process of combination of the seeds, and this, like a bank safe, can only be opened by two keys in combination. The principle of life does not belong to one, but to a combination.

Take any plot of wild land. We will consider the reproduction and decay there for a period of x years.

Say there are k seeds to start with, and s seeds are blown in or out each year by chance.

And b seeds reproduce and produce other seeds each year, and g of these are preserved.

Hence the total number of seeds at the end of x years is $k + xs + ge^{bx} = k + xs + gc^x$, by Proposition 2 and Definition 14.

And these are the things from which the combinations are made.

Hence if the total number of plants at the end of x years is P, $k + xs + gc^x = \log P$, since the logs are the things from which the combinations are made.

Now if we count the number of plants in each of 4 years, t years apart, we get 4 equations from which k, s, g and c may be found.

And as t may have any value and may cover different periods we will get slightly different values for these constants depending on the periods, owing to the fluctuations of chance.

This formula agrees very closely with the Mortality Tables used by Life Insurance Companies.

It is useless to form normal equations from this general *obs* equation because they could not be solved.

But if we assume a value for c we can make as many *obs* equations as we like, and can form normal equations from which we can find the most likely values of k, s and g for this value of c.

The *Institute of Actuaries Text-Book*, Part II., p. 83, gives $c = 1{\cdot}0956122$ for the Mortality Tables used.

If this is not the most likely value of c, and it certainly is not, the process may be repeated with different trial values of c, having regard to the weights, until we get the most likely values of all the four constants, so that the results will agree very closely with the actual Mortality, by Definition 14.

The increase in population of any country follows the same law, and the same formula can be used, by finding new values for the four constants.

From the populations at four different equi-distant dates an approximate value of c may be found.

And using this value of c the most likely values of k, s and g may be found for this value of c.

Then using slightly different trial values of c, having regard to the weights, the most likely values of all the constants can be found, by Definition 14.

Example 528. From a Table of decennial populations of England find the most likely population in 1828, and the chance that it would be 100,000 more than it was. And find the dates of any great crises in the history of England as shown by abnormal increase or decrease of population. And the effect of each on the life of England?

Example 529. Any Table of Statistics has some general formula, and the most likely constants can be found. After all known causes have been eliminated the remaining fluctuations are due to chance, by Definition 1. The variations in the values of the constants are due to chance, if we have no knowledge as to their cause, by Definition 1.

Example 530. If anything is *obs* twice under the same set of conditions the difference should be 0, because if anything is measured in one direction and then back again they finish at the same point from which they started.

Hence the discrepancy d is a true error, and not a residual error. And if we have no other information it must be taken as the normal error, by Definition 12.

Hence the normal error for one *obs* is $\sqrt{\frac{d^2}{2}} = \frac{d}{\sqrt{2}}$, if the weight of a single *obs* is 1.

And if the weight of a single *obs* is p the normal error of an *obs* of weight 1 is $\sqrt{\frac{pd^2}{2}}$, by Definition 13.

And if n things are each *obs* twice the normal error for an *obs* of weight 1 is

$$x_0 = \sqrt{\frac{p_1d_1^2 + p_2d_2^2 + p_3d_3^2 + \text{etc.}}{2n}} = \sqrt{\frac{\Sigma pd^2}{2n}}.$$

And the Median error of an *obs* of weight 1 is

$$r = \cdot 6745 \sqrt{\frac{\Sigma pd^2}{2n}} = \cdot 4769 \sqrt{\frac{\Sigma pd^2}{n}}.$$

Thus by measuring a few lines in any survey twice, a surveyor can find the Median error for a line of weight 1 of his chainmen, without any waste of time.

The Median error for an angle can be found in the same way.

And this information should be entered in the notes so as to have a record of the precision with which the work was done, as well as temperature, instruments, etc.

All constant errors must be removed as far as possible, for they are eliminated in this operation.

The weight of a line is inversely proportional to its length if each part is measured under the same set of conditions, because the number of chances of a pair of contradictories is proportional to the length if each part is measured under the same set of conditions, by Proposition 11 and Definition 13.

Example 531. If the lengths got for four lines under the same set of conditions are

1001·03	and	1001·06
647·03	and	647·09
846·51	and	846·65
725·47	and	725·39

what is the Median error of a line of length $l_1 = 100$?

Example 532. If A, B and C each *obs* anything and get the values a, b and c, respectively, what is its most likely value if we have no other knowledge such as that found in the last Example ? What is the chance that the error of the most likely value is greater than k ? What is the chance that the error of b is greater than k ?

If d is the discrepancy between any two *obs* we have

$$a - b = d_{ab}$$
$$a - c = d_{ac}$$
$$b - c = d_{bc}.$$

The number of these *obs* equations is $\frac{\lfloor n}{\lfloor 2 \lfloor n-2}$, if there are n observers, by Proposition 2.

Let the most likely Normal errors of a, b and c be x_a, x_b and x_c, respectively, and the weights, when unknown, of a single *obs* be 1.

The most likely value of anything changes as our knowledge is increased, by Def. 1, as in Prop. 6 and Example 186, which give the most likely, but not the true, values.

Any equation is a single indirect *obs* if it has not been directly *obs*. Hence $a - b = d_{ab}$ is a single indirect *obs*. Hence its weight is 1. But a is a single direct *obs*. Hence its weight is 1. Hence the weight of a is equal to the weight of $a - b$. Hence $x_a^2 = d_{ab}^2$, by Def. 13.

Again $a - c = d_{ac}$. Hence $x_a^2 = d_{ac}^2$. Hence with all our knowledge the most likely value of x_a^2 is $\frac{1}{2}(d_{ab}^2 + d_{ac}^2)$. Similarly $x_b^2 = \frac{1}{2}(d_{ab}^2 + d_{bc}^2)$ and $x_c^2 = \frac{1}{2}(d_{ac}^2 + d_{bc}^2)$.

We can get the same result, subject to the vagaries of chance, as in Example 186, taking $a + b$, etc., instead of $a - b$, etc. For if we have only the equation $a - b = d_{ab}$ to find x_a we must assume that there is no error in b, because we know nothing about the error in b except that its most likely value is 0, by Prop. 12. Any other assumption is certainly false. Hence this one is true.

See remarks about disagreement between the Normal and average errors on page 64. The same thing with different weights and different things with the same weight are alike.

And if X is the normal error of the most likely value, or weighted average, we have

$$\frac{1}{X^2} = \frac{1}{x_a^2} + \frac{1}{x_b^2} + \frac{1}{x_c^2}$$

to find X.

The most likely value, or weighted average, is

$$\frac{p_a a + p_b b + p_c c}{p_a + p_b + p_c}.$$

The normal error for an *obs* of weight 1 is

$$\sqrt{p_a x_a^2} = \sqrt{p_b x_b^2} = \sqrt{p_c x_c^2}.$$

And similarly for any number of observers.

Example **533**. If A and B measure a line and get l_{ab}, and B and C measure it and get l_{bc}, and A and C measure it and get l_{ac}, what is the most likely length, etc.?

$$l_{ab} - l_{ac} = d_1, \text{ etc.}$$

In the first equation the weight of $2x_a^2$ is 2. Hence $2x_a^2 = d_1^2$ must be taken twice. $x_a^2 = \frac{1}{6}(2d_1^2 + d_2^2 + d_3^2)$, $x_b^2 = \frac{1}{6}(d_1^2 + 2d_2^2 + d_3^2)$ and $x_c^2 = \frac{1}{6}(d_1^2 + d_2^2 + 2d_3^2)$.

Example **534.** If A *obs* anything and gets **1,000**, and B *obs* it and gets **1,002**, and C *obs* it and gets **1,002·1**, what are the respective weights of the three *obs* ? What is the most likely value ? What is the chance that it is less than **999** ? What is the chance that it is more than **1003·1** ?

Example **535.** If A, B, C and D *obs* it and get the values **1,000**, **1,002**, **1,002·1** and **1,002·2**, what are the respective weights ? What is the most likely value ? What is the chance that it will be less than **999** ? What is the chance that it will be more than **1,003·1** ? What is the chance that it will be more than **1,006** ?

Example **536.** Mark Wright's Foot-stool in the Grotto at Edmonton, Alberta, has been *obs* six times by six lines of levels to Frank Ford's Bench, and the elevations of the former above mean sea level in feet were

2180·1
2179·8
2179·9
2180·3
2180·0
2174·6.

What are the respective weights of the *obs* ? What is the most likely elevation ? What is the chance that it is less than **2174·6** feet ? What is the most likely normal error of the last obs ? Of the first ? Of the weighted average ? What is the Median elevation ?

Before the last *obs* was made what were the respective weights? What was the most likely elevation? What was the most likely normal error of the first *obs*? Of the weighted average? What was the Median elevation? What was the chance that the elevation is less than 2,174·6 feet? That some time the error was over 10 ft?

There is no information as to the sets of conditions under which the *obs* were made. If they were all made under the same set of conditions, what is the most likely elevation?

Example 537. Show that any variable that travels far without supplies must go in harmonic waves.

Anything that travels far without supplies must make a lot of motion without expending much energy.

Hence its force must not move in the direction of itself, but at right angles to it, very nearly, where it does no work.

A force represented by the radius and acting on the centre of a circle may revolve around that centre without doing any work. And at a terrific rate.

It does no work, but its rectangular components in any two given directions produce harmonic motion as the radius revolves. Space is full of pulsations.

And this harmonic motion in an elastic medium constitutes the harmonic waves of electricity, light, heat, sound, the sea and the revolution of planets.

They travel far but do no work, or very little, because it is a constant give and take.

When the revolutions are very rapid the ultimate elements only of the medium are acted on. Hence when this rapidity is reached all media give nearly the same velocity, and are called ether. Hence no waves travel faster than light, electricity, etc.

This is why wireless telegraphy travels so far, and so fast. Little time for motion, only force.

What is the resemblance, and the difference, between this and the law of error ?

Why is the law of error the same as that of evolution ? Because they are both caused by the interference of yes and no, as time goes on. Error is a resultant of two lines at right angles to each other, and not a simple sum like the expectation. It is geometry, since one of its factors is the angle through which it can turn.

Example 538. What is the chance that during 10,000 flips of a coin the error will never be greater than 600 ?

The most likely error is $x_0 = 50$.

The Median error at the end of the series is

$$r = \cdot 6745 \text{ of } 50 = \tfrac{2}{3} \text{ of } 50 = 33\cdot 3, \text{ approximately.}$$

$$x = 600 = \frac{600}{33\cdot 3}\, r = 18r = sr.$$

$$(\tfrac{1}{2})^{\frac{s^2}{3}} = (\tfrac{1}{2})^{\frac{(18)^2}{3}} = (\tfrac{1}{2})^{108}.$$

$$\text{Log}(\tfrac{1}{2})^{108} = -\cdot 30103 \times 108 = -32.51 = \overline{33}.49.$$

Hence $(\tfrac{1}{2})^{\frac{s^2}{3}}$ is decimal 32 zeros followed by 3.

Which is the chance that the error will be greater than 600 some time during the series.

Hence the chance that it will not be greater than 600 at any time during the series is 1 minus this decimal, by Definition 3.

The chance that the error will not be greater than 600 at the end of the series is the same, very nearly, by Example 517.

Example 539. What is the chance that the error will never be greater than $x = 3$ during the series?

$$\frac{x}{r} = \frac{3}{33{\cdot}3} = \frac{1}{11}\text{, approximately.}$$

Hence $x = \frac{1}{11} r = sr$.

$$(\tfrac{1}{2})^{\frac{s^2}{3}} = (\tfrac{1}{2})^{\frac{1}{363}} = \frac{3 - \frac{1}{363}}{3 + \frac{1}{363}} = \frac{1089 - 1}{1089 + 1} = \frac{1088}{1090}$$

$$= \frac{1088 - 90}{1090 - 90} = \frac{998}{1000} = {\cdot}998\text{, approx., by Example 27.}$$

Hence the chance that the error will never be greater than 3 during the series is

$$1 - {\cdot}998 = {\cdot}002\text{, approx.}$$

Or we can use the formula $e^{-\frac{s^2}{4{\cdot}4}}$, by Example 514.

$$e^{-\frac{s^2}{4{\cdot}4}} = e^{-\frac{1}{532}} = \frac{2 - \frac{1}{532}}{2 + \frac{1}{532}} = \frac{1063}{1065} = \frac{1063 - 65}{1065 - 65}$$

$= {\cdot}998$, by Example 26.

This error is so much less than the most likely error that the result is not precise.

Example 540. What is the chance that during 10,000 flips of a coin the error will never be more than 4 times the most likely error?

The most likely error is $x_0 = 50$.

The chance that the error will be greater than x some time during the series is $e^{-\frac{x^2}{2x_0^2}}$.

$$e^{-\frac{x^2}{2x_0^2}} = e^{-\frac{m^2x_0^2}{2x_0^2}} = e^{-\frac{m^2}{2}} = (\tfrac{1}{2})^{\frac{m^2}{2\log^e 2}}$$

$$= (\tfrac{1}{2})^{\frac{m^2}{2 \times \cdot 69315}} = (\tfrac{1}{2})^{\frac{m^2}{1\cdot 4}} = (\tfrac{1}{2})^{\frac{16}{1\cdot 4}} = (\tfrac{1}{2})^{11\cdot 4}$$

very nearly.

$\text{Log}(\tfrac{1}{2})^{11\cdot 4} = -\cdot 30103 \times 11\cdot 4 = -3\cdot 43 = \bar{4}\cdot 57.$

Hence $(\tfrac{1}{2})^{11\cdot 4} = \cdot 0004.$

Hence the chance that the error will never be greater than $4x_0$ is $1 - \cdot 0004 = \cdot 9996.$

The chance that the error will never be greater than mx_0 in any series is $1 - (\tfrac{1}{2})^{\frac{m^2}{1\cdot 4}}$, very nearly.

Example 541 What is the chance that the error at the end of any series will be more than 12 times the Median error r?

It is $(\tfrac{1}{2})^{\frac{s^2}{3}}$.

$$\text{Log}\,(\tfrac{1}{2})^{\frac{s^2}{3}} = -\frac{s^2}{3}\log 2 = \overline{15}\cdot 55.$$

Hence the chance is decimal 14 zeros followed by 4. See Example 517.

Example 542. What is the chance that the error at the end of any series will be more than $7r$?

It is $\frac{1}{5}(\tfrac{1}{2})^{\frac{s^2}{3}}$, very nearly.

$$\text{Log}\,(\tfrac{1}{2})^{\frac{s^2}{3}} = -\frac{s^2}{3}\log 2 = \bar{5}\cdot 08.$$

Hence the chance is $\frac{1}{5}$ of ·000,01 = ·000,002, very nearly.

Example 543. What is the chance that the error at the end of any series will be more than + 7*r* ?

It is $\frac{1}{10}\left(\frac{1}{2}\right)^{\frac{s^2}{3}}$ = ·000,001, very nearly.

Example 544. What is the chance that the error at the end of any series will be more than 3*r* ?

$$\text{It is } \frac{\left(\frac{1}{2}\right)^{\frac{s+1}{2}}}{\underline{|s}} = \frac{1}{24} = \cdot 042, \text{ very nearly, by Prop. 7.}$$

Example 545. What is the chance that the error at the end of any series will be more than 2*r* ?

$$\text{It is } \frac{\left(\frac{1}{2}\right)^{\frac{s+1}{2}}}{\underline{|s}} = \cdot 707\,\frac{\left(\frac{1}{2}\right)^{\frac{s}{2}}}{\underline{|s}} = \cdot 177, \text{ by Prop. 7.}$$

Example 546. If to break one's word is an error of 50 and to steal is an error of 100, and a man's Median error is 10, what is the chance that he would break his word at the end of the temptation ? What is the chance that he would steal at the end of the temptation ?

To break one's word is $5r = sr$, and to steal is $10r = sr$.

$$t = \frac{x}{x_0\sqrt{2}} = \frac{sr}{2\cdot 1r} = \frac{s}{2\cdot 1}.$$

Hence $t_1 = 2\cdot 38$.

$t_2 = 4\cdot 76$.

Hence $\frac{1 - P_1}{2} = \; = \cdot 00038$.

And $\frac{1 - P_2}{2}$ = ·000,000,000,01, by the Tables in the back of the book.

Example 547. What is the chance that some time during the temptation he would steal?

It is $\frac{1}{2}\left(\frac{1}{2}\right)^{\frac{s^2}{3}}$ = ·000,000,000,05, approx.

Example 548. This book is the study of the habits of things, and shows how to find the chance of any variation from the habits, that anything rarely departs far from its habits, and that a parent's duty is to see that children form good habits, and to follow their example. Habits of virtue, industry, patience and courtesy cause success and happiness. And when these habits are formed they are not only easy to follow but hard to depart from.

Any thing has habits whether we know their causes or not. We can know the nature of a thing by its habits almost as well as if we knew the causes of its actions. We cannot know what it will do next, for the cause of any action is often composed of a very large number of varying elements. Each separate action is erratic and seems to have no cause.

We cannot know the nature of a thing from one or two of its actions. These may be as different as night and day. But if we watch it a while we shall learn its habits, and its nature, without knowing anything about the causes of its actions, for it seldom departs far from its habits.

And when we know the nature of a thing we know what the general course of its habits will be in the future, without knowing the cause of any of its actions.

The course of the habits of anything is called the

expectation, and any variation from this is called an error.

Nearly all the knowledge that we have is based on this subject. We know little for certain, for most of our knowledge is gained from experience, and anything may happen any number of times in succession by pure chance.

And since this constantly occurs in different degrees in life, and to all, everyone should know how to find the chance that his daily observations will not differ more than a given amount from future experience.

It provides a scale by which he can, in a moment and mentally, weigh all his observations, and prophesy from them.

By shooting a few shots at a target a soldier can find his median error. And when he has found it all his subsequent experience, unless he becomes more expert, will conform so closely to it that, when he first sees this, it will almost frighten him to find that so many erratic shots when counted up hardly change his label by a difference large enough to be measured. It seems to him as if the witches are at work. He can tell just how many shots he should put in any ring, no matter how many rings there are, out of 100 shots, and will do almost exactly this in practice.

And this subject applies to every part of life with the same precision as to shooting in war. The results are most marvellous, and the witches of old become real, and prophesy for us in all phases of life, and tell what happened in the old days before history began.

Example 549. And in things that produce offspring which again produce other offspring the variations become permanent and the original type is largely lost in a very large number of generations. And as the environment is more favourable to some variations than to others these become predominant while some become extinct. And as the environment gradually changes these predominant variations fluctuate and gradually become extinct in their turn. And the variations gradually rise in type as it is the favoured variations that survive. Thus the fosils in the successive geological formations rise in type, fluctuate and become extinct, but the favoured variations give a gradual rise in type although the predominant ones in any geological period become extinct, for their offspring that are favoured by the new conditions continue to rise in type. And the law of Evolution is the same as that of error, the interference of yes and no. If yes and no did not interfere there would be no error, and we would always have the expectation or true value, by Definition 11, and there would be no variation of species. The original type would always remain just the same, and the life in the earliest geological formations would be identical with that in the world to-day.

Thus we see that the law that makes and loses fortunes at Monte Carlo is the same as that of Evolution, the interference of contradictories.

Example 550. The interference of contradictories makes abnormal combinations of the elements, whether in a game of chance, in measuring anything, in all the actions of men and women, in the growth of a tree or in the transformation of the elements of an egg as it passes through its ancestral forms. The copy is seldom just the same as the model, or habit, and the variation of chance is the error.

TABLES OF t AND P

t	P
·00	·0000000
·01	·0112833
·02	·0225644
·03	·0338410
·04	·0451109
·05	·0563718
·06	·0676215
·07	·0788577
·08	·0900781
·09	·1012806
·10	·1124630
·11	·1236230
·12	·1347584
·13	·1458671
·14	·1569470
·15	·1679959
·16	·1790117
·17	·1899923
·18	·2009357
·19	·2118398
·20	·2227025
·21	·2335218
·22	·2442958
·23	·2550225
·24	·2657000
·25	·2763263
·26	·2868997
·27	·2974182
·28	·3078800
·29	·3182834
·30	·3286267
·31	·3389081
·32	·3491259
·33	·3592785
·34	·3693644
·35	·3793819
·36	·3893296
·37	·3992059
·38	·4090093
·39	·4187385
·40	·4283922
·41	·4379690
·42	·4474676
·43	·4568867
·44	·4662251
·45	·4754818
·46	·4846555
·47	·4937452
·48	·5027498
·49	·5116683
·50	·5204999
·51	·5292437
·52	·5378987
·53	·5464641
·54	·5549392
·55	·5633233
·56	·5716157
·57	·5798158
·58	·5879229
·59	·5959365
·60	·6038561
·61	·6116812
·62	·6194114
·63	·6270463
·64	·6345857
·65	·6420292

t	P	t	P
·66	·6493765	1·05	·8624360
·67	·6566275	1·06	·8661435
·68	·6637820	1·07	·8697732
·69	·6708399	1·08	·8733261
·70	·6778010	1·09	·8768030
·71	·6846654	1·10	·8802050
·72	·6914330	1·11	·8835330
·73	·6981038	1·12	·8867879
·74	·7046780	1·13	·8899707
·75	·7111556	1·14	·8930823
·76	·7175367	1·15	·8961238
·77	·7238216	1·16	·8990862
·78	·7300104	1·17	·9020004
·79	·7361035	1·18	·9048374
·80	·7421010	1·19	·9076083
·81	·7480033	1·20	·9103140
·82	·7538108	1·21	·9129555
·83	·7595238	1·22	·9155339
·84	·7651427	1·23	·9180501
·85	·7706680	1·24	·9205052
·86	·7761002	1·25	·9229001
·87	·7814398	1·26	·9252359
·88	·7866873	1·27	·9275136
·89	·7918432	1·28	·9297342
·90	·7969082	1·29	·9318987
·91	·8018828	1·30	·9340080
·92	·8067677	1·31	·9360632
·93	·8115635	1·32	·9380652
·94	·8162710	1·33	·9400150
·95	·8208908	1·34	.9419137
·96	·8254236	1·35	·9437622
·97	·8298703	1·36	·9455614
·98	·8342315	1·37	·9473124
·99	·8385081	1·38	·9490160
1·00	·8427008	1·39	·9506733
1·01	·8468105	1·40	·9522851
1·02	·8508380	1·41	·9538524
1·03	·8547842	1·42	·9553762
1·04	·8586499	1·43	·9568573

t	P	*t*	P
1·44	·9582966	1·83	·9903467
1·45	·9596950	1·84	·9907359
1·46	·9610535	1·85	·9911110
1·47	·9623729	1·86	·9914725
1·48	·9636541	1·87	·9918207
1·49	·9648979	1·88	·9921562
1·50	·9661052	1·89	·9924793
1·51	·9672768	1·90	·9927904
1·52	.9684135	1·91	·9930899
1·53	·9695162	1·92	·9933782
1·54	·9705857	1·93	·9936557
1·55	·9716227	1·94	·9939229
1·56	·9726281	1·95	·9941794
1·57	·9736026	1·96	·9944263
1·58	·9745470	1·97	·9946637
1·59	·9754620	1·98	·9948920
1·60	·9763484	1·99	·9951114
1·61	·9772069	2·00	·9953223
1·62	·9780381	2·01	·9955248
1·63	·9788429	2·02	·9957195
1·64	·9796218	2·03	·9959063
1·65	·9803756	2·04	·9960858
1·66	·9811049	2·05	·9962581
1·67	·9818104	2·06	·9964235
1·68	·9824928	2·07	·9965822
1·69	·9831526	2·08	·9967344
1·70	·9837904	2·09	·9968805
1·71	·9844070	2·10	·9970205
1·72	·9850028	2·11	·9971548
1·73	·9855785	2·12	·9972836
1·74	·9861346	2·13	·9974070
1·75	·9866717	2·14	·9975253
1·76	·9871903	2·15	·9976386
1·77	·9876910	2·16	·9977472
1·78	·9881742	2·17	·9978511
1·79	·9886406	2·18	·9979505
1·80	·9890905	2·19	·9980459
1·81	·9895245	2·20	·9981372
1·82	.9899431	2·21	·9982244

t	P	*t*	P
2·22	·9983079	2·61	·9997767
2·23	·9983878	2·62	·9997888
2·24	·9984642	2·63	·9998003
2·25	·9985373	2·64	·9998112
2·26	·9986071	2·65	·9998215
2·27	·9986739	2·66	·9998313
2·28	·9987377	2·67	·9998406
2·29	·9987986	2·68	·9998494
2·30	·9988568	2·69	·9998578
2·31	·9989124	2·70	·9998657
2·32	·9989655	2·71	·9998732
2·33	·9990162	2·72	·9998803
2·34	·9990646	2·73	·9998870
2·35	·9991107	2·74	·9998933
2·36	·9991548	2·75	·9998994
2·37	·9991968	2·76	·9999051
2·38	·9992369	2·77	·9999105
2·39	·9992751	2·78	·9999156
2·40	·9993115	2·79	·9999204
2·41	·9993462	2·80	·9999250
2·42	·9993793	2·81	·9999293
2·43	·9994108	2·82	·9999334
2·44	·9994408	2·83	.9999372
2·45	·9994694	2·84	·9999409
2·46	·9994966	2·85	·9999443
2·47	·9995226	2·86	·9999476
2·48	·9995472	2·87	·9999507
2·49	·9995707	2·88	·9999536
2·50	·9995930	2·89	·9999563
2·51	·9996143	2·90	·9999589
2·52	·9996345	2·91	·9999613
2·53	·9996537	2·92	·9999636
2·54	·9996720	2·93	·9999658
2·55	·9996893	2·94	·9999679
2·56	·9997058	2·95	·9999698
2·57	·9997215	2·96	·9999716
2·58	·9997364	2·97	·9999733
2·59	·9997505	2·98	·9999750
2·60	·9997640	2·99	·9999765

t	P	*t*	P
3·00	·9999779	3·39	·99999836842
3·01	·9999793	3·40	·99999847899
3·02	·9999805	3·41	·99999858255
3·03	·9999817	3·42	·99999867947
3·04	·9999829	3·43	·99999877011
3·05	·9999839	3·44	·99999885482
3·06	·9999849	3·45	·99999893394
3·07	·9999859	3·46	·99999900780
3·08	·9999867	3·47	·99999907672
3·09	·9999876	3·48	·99999914101
3·10	·9999884	3·49	·99999920097
3·11	·9999891	3·50	·99999925691
3·12	·9999898	3·51	·99999930905
3·13	·9999904	3·52	·99999935766
3·14	·9999910	3·53	·99999940296
3·15	·9999916	3·54	·99999944519
3·16	·9999921	3·55	·99999948452
3·17	·9999926	3·56	·99999952115
3·18	·9999931	3·57	·99999955527
3·19	·9999936	3·58	·99999958703
3·20	·9999940	3·59	·99999961661
3·21	·9999944	3·60	·99999964414
3·22	·9999947	3·61	·99999966975
3·23	·9999951	3·62	·99999969358
3·24	·9999954	3·63	·99999971574
3·25	·9999957	3·64	·99999973636
3·26	·9999960	3·65	·99999975551
3·27	·9999962	3·66	·99999977333
3·28	·9999965	3·67	·99999978990
3·29	·9999967	3·68	·99999980528
3·30	·9999969	3·69	·99999981957
3·31	·9999971	3·70	·99999983285
3·32	·9999973	3·71	·99999984517
3·33	·9999975	3·72	·99999985663
3·34	·9999977	3·73	·99999986726
3·35	·9999978	3·74	·99999987712
3·36	·9999980	3·75	·99999988629
3·37	·9999981	3·76	·99999989477
3·38	·9999982	3·77	·99999990265

t	P	t	P
3·78	·99999990995	3·94	·99999997482
3·79	·99999991672	3·95	·99999997678
3·80	·99999992200	3·96	·99999997860
3·81	·99999992881	3·97	·99999998028
3·82	·99999993421	3·98	·99999998183
3·83	·99999993921	3·99	·99999998327
3·84	·99999994383	4·00	·99999998459
3·85	·99999994812	4·10	·99999999330
3·86	·99999995208	4·20	·99999999714
3·87	·99999995575	4·30	·99999999880
3·88	·99999995915	4·40	·99999999951
3·89	·99999996230	4·50	·99999999981
3·90	·99999996522	4·60	·99999999992
3·91	·99999996790	4·70	·99999999997
3·92	·99999997039	4·80	·99999999999
3·93	·99999997260		

INDEX

For Product Safety Concerns and Information please contact our EU representative GPSR@taylorandfrancis.com Taylor & Francis Verlag GmbH, Kaufingerstraße 24, 80331 München, Germany

Batch number: 10725859

Printed by Printforce, the Netherlands